'Here Is the Weather Forecast'

'Here is the weather forecast till 6 pm tomorrow. An anticyclone is centred over Denmark. This is likely to persist. In the Atlantic a depression is moving slowly northwards from the Azores towards Iceland. Tonight skies will be mainly clear and air temperatures will fall to around 4° Centigrade over much of England, Scotland and Wales. In a few sheltered places, especially in the east, there may be a slight frost. After early morning mist, tomorrow will be clear and sunny. Temperatures will rise to around 12° Centigrade in the south and 10° Centigrade in the north. Later in the day western districts will become cloudy. The outlook until the weekend is one of little change.'

The weather forecaster can be seen on television making his or her forecast. The forecaster gets the information from a number of *weather stations* which are set up over a

A television weather forecaster at work

A weather map for a day in early October

wide area. The map on the left shows all the information supplied by a number of these stations at the same moment in time. It is called a *synoptic chart*. It provided the information for the forecast given. Information recorded over a period of time, such as sunshine and rainfall, is not shown.

Each station is shown on the map by a *station circle*. The amount of cloud is shown by the symbol inside the circle. The direction and speed of the wind are shown by an arrow with a number of tails on it. The air temperature in degrees Centigrade is written beside the circle, with a symbol for present weather beneath it.

The map has a lot of lines drawn on it. These are called *isobars*. They join places which have the same *pressure* (weight of air) at any moment in time. Together they are showing pressure patterns. The numbers written beside the lines such as 1020 or 996, show the pressure measured in *millibars*.

On the map two lines called *fronts* are cutting across the isobars. The line with black semicircles along one side of it represents a *warm front*. The line with black triangles along it marks a *cold front*. These lines show where two air streams meet.

A lot of work has to be done quickly before weather maps can be drawn. The weather man, or *meteorologist*, then studies the maps carefully and makes the forecast. This book shows how meteorologists collect information about the weather to make their maps.

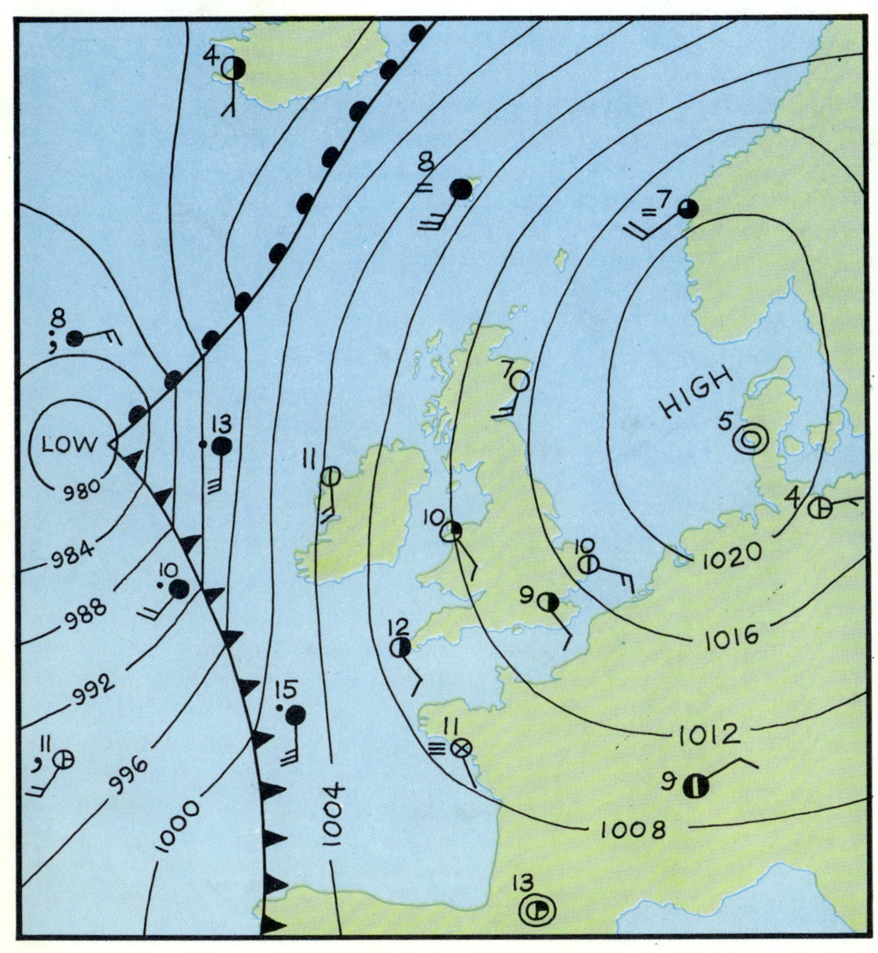

1

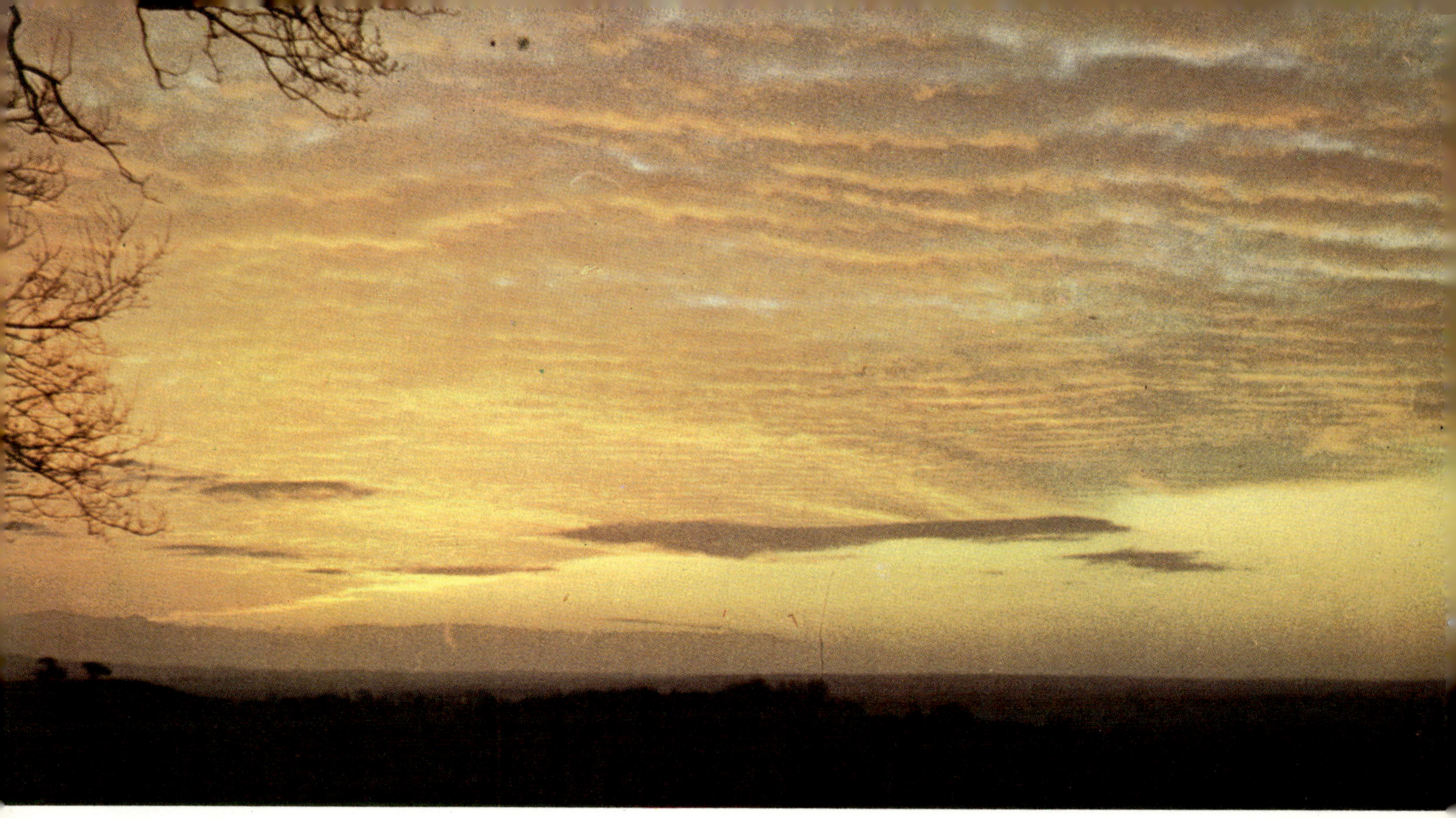

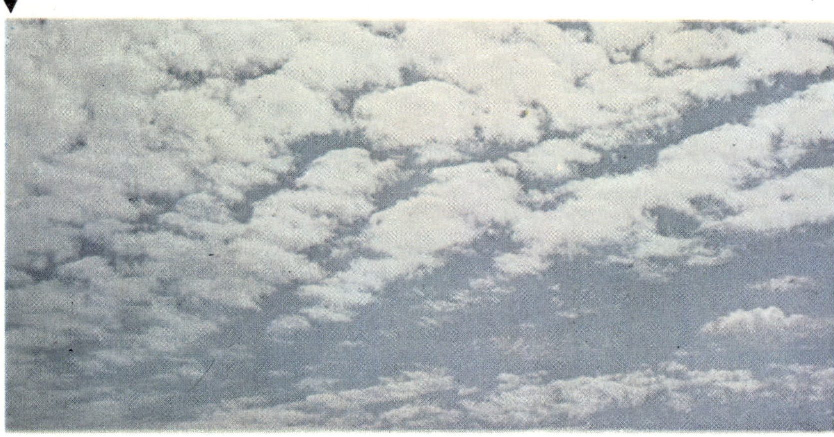

Weather Sayings and Superstitions

Above: 'Red sky at night, shepherd's delight; red sky in the morning, shepherd's warning'

'Mackerel sky, mackerel sky, not long wet and not long dry'
▼

Man has always been concerned with his weather. From early times all kinds of weather sayings have been made up. Some are still used today.

If a saying is based on legend or superstition it has little scientific value. For example, in England it is said that the weather on July 15th, known as St Swithin's Day, always continues for the next 40 days. Examinations of the records show that this is not true. No matter what the weather on St Swithin's Day, the period from July 16th to August 24th has an *average* of 24 dry days and 16 wet days in lowland Britain.

The British also have the saying 'Mackerel sky, mackerel sky, not long wet and not long dry'. Mackerel sky is a pattern of small, rounded clouds strung in lines across the sky. It looks like the scaley back of that fish. Mackerel sky is one of the cloud types that can be seen before a period of rain at a warm front (see page 22, bottom diagram).

Clear skies at night may lead to a rapid drop in temperature (see page 14). The saying which describes this is 'Clear sky, frost soon'.

'Red sky at night, shepherd's delight; Red sky in the morning, shepherd's warning'. The redness of the sky is explained by the overall pattern of the clouds in the sky, and the position of the sun. If the sky is fairly clear at sunset, the sun's rays may strike the undersides of any clouds in the eastern sky or overhead. The rays pass at a low angle through a great thickness of the atmosphere and, as a result, give the clouds a reddish colour.

The first line of the saying refers to the weather in the west at sunset. In latitudes where the weather moves from west to east the saying means that it may be a clear day tomorrow.

When the cloud is building up in the west but the eastern sky is clear, the second line of the saying applies at sunrise.

1. Test the weather sayings quoted on this page and see how often they are correct.
2. Collect other weather sayings and test their accuracy.

Can Man Change the Weather?

Two people living in the same place may want completely different weather at the same moment. They cannot both be satisfied with the weather which actually takes place. Suppose that they agreed that they wanted a whole month of clear blue skies and high temperatures. Could anything be done to make this happen?

1. Make a list of some special weather conditions the following might prefer: farmers, yachtsmen, holiday-makers, glider pilots.

Believing that they could influence the weather, early peoples often made sacrifices to persuade their gods to send rain for their withering crops in times of drought. But weather experts know that man is unable to influence his weather over large areas of land. The amount of energy which is involved

An aerial view of wind breaks in the lower Rhône valley in Southern France. The Mistral wind blows here from north to south. From the positions of the wind breaks, it is possible to turn this picture to find the direction of the Mistral.

The lawns behind the White House, Washington, being sprayed during a spell of dry weather

Greenhouses create artificial temperatures. Flowers and vegetables can be grown in them out of season.

in one small thunderstorm is much more than that released by the largest hydrogen bomb yet exploded. So, for the moment, large-scale weather management is impossible.

On a small scale man can change his weather. He can water his garden during a dry spell or turn on the air-conditioning in his house during a heat wave. During World War II (1939–45) fog sometimes made it difficult for aircraft to land at their bases after night bombing raids. Runways were then surrounded with pots of burning oil. In this way the surrounding air was heated and the fog dispersed.

In some parts of the world strong, cold winds can blow for a short time during the warm season. These winds can damage valuable crops. To prevent this wind breaks are set up. In the Rhône valley in southern France, poplar trees are grown to protect crops against the *Mistral* wind.

A number of experiments have been carried out to make rain by *cloud seeding*. A cloud has to be found which could produce rain if conditions were right. Then chemicals have to be added to the cloud. Sometimes solid carbon dioxide or silver iodide is dropped from an aircraft. In this way some rain has been made, but the method is not highly reliable. It is also costly.

2. In what ways is it possible to 'change' the weather in your own garden?

3. Which is more dangerous: an unpleasant weather feature which you know may happen a few times each year, and for which you can prepare; or an unpleasant weather feature which may happen once in ten years? Explain your answer, and give examples of each.

The Earth's Atmosphere

All of the winds, clouds, pressure patterns, and other weather features which affect man's life are found in the bottom layer of the atmosphere that surrounds the earth. This layer is called the *troposphere*. It is only about 16 kilometres thick above the equator and about 8 kilometres thick above the poles. The air in the troposphere contains 78% nitrogen, 21% oxygen and 1% other gases. The other gases include a small but important amount of water vapour.

The tall diagram on the right shows the lowest parts of the earth's atmosphere. The smaller diagram at the foot of this column shows the troposhere in greater detail. Both diagrams show the height of things that can be seen in the sky, such as clouds and the highest aeroplanes.

1. Which are the warmest layers in the lowest 110 kilometres of the earth's atmosphere?

2. What change in temperature would a thermometer attached to a balloon record as it rises through the troposphere?

If the sun did not shine on the earth's surface there would be no weather. The sun sends out, or *transmits*, energy in the form of short-wave rays in all directions into space. When some of these reach the earth they are absorbed by its surface and other solid objects. Each of these objects then *radiates* (gives out) heat rays into the troposphere. As a result, the air closest to the ground is usually warmer than the air above it.

3. The amount of the sun's energy in the form of short-wave rays which reaches the earth's surface varies from time to time. List some of the weather conditions which can affect air temperature. How is air temperature also affected by time of day and season of the year?

A section view of part of the earth's atmosphere showing pressures and temperatures at different heights. The column on the far right shows pressures. The white line zig-zagging through the diagram is a temperature graph. Its scale is at the bottom of the diagram. The bands of red and blue suggest the warm and cold zones of the atmosphere.

The small diagram below shows the position of the tropopause at the bottom of the atmosphere. It is below the tropopause that all of the earth's weather is found. The higher layers of the atmosphere, such as the stratosphere and the mesosphere, are of little concern to weather men.

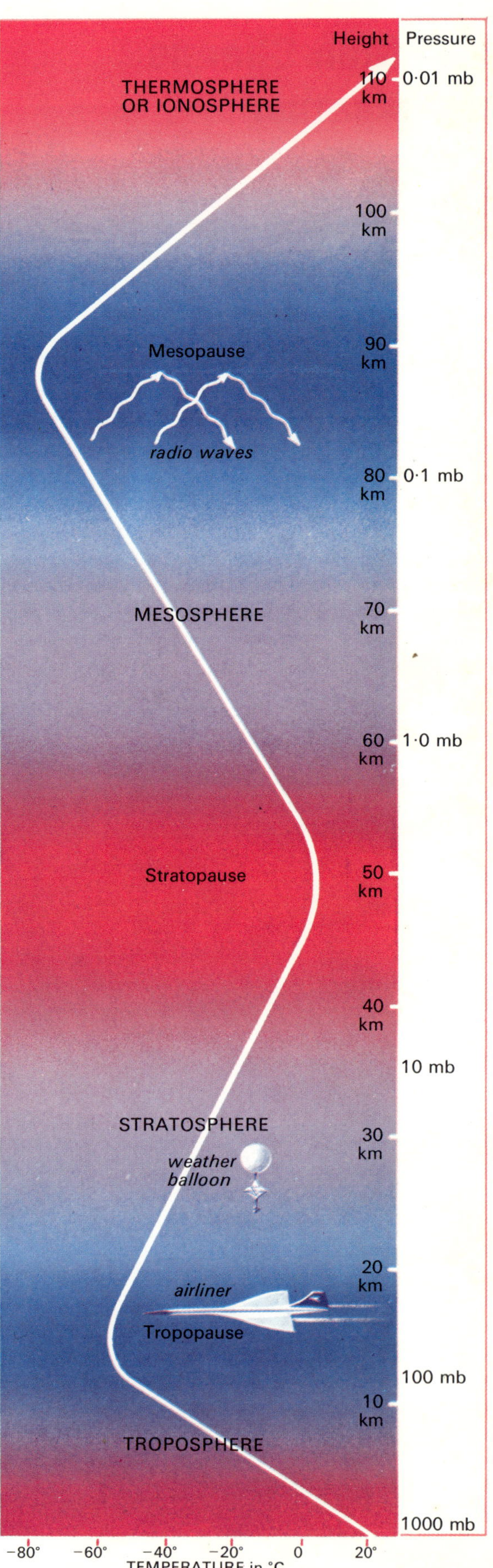

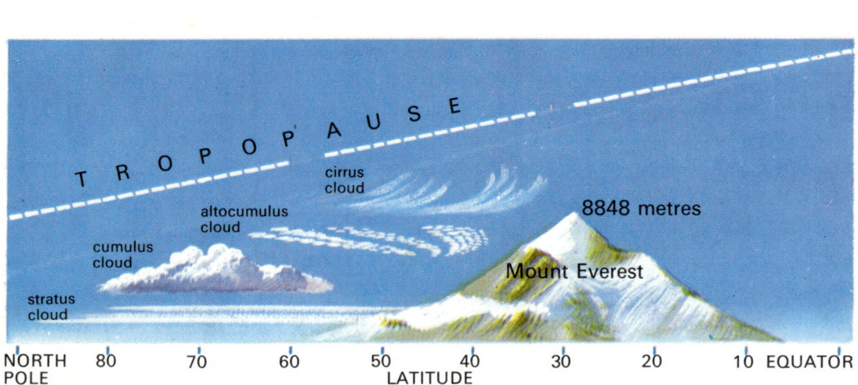

Air movements in the troposphere

Differences in air temperatures can cause differences in air pressure. The air is hottest where most energy is received from the sun. This is usually in tropical latitudes where the sun appears highest in the sky. Below is a section view which shows how the air moves about in the troposphere. The global map on the right shows the movements of air near to the earth's surface. These movements of the air are, in fact, the earth's main wind belts. They are called *planetary winds*.

Look at both diagrams. Heated air rises above the equator (1). At ground level there is then an area of low pressure (2). When the air reaches the tropopause it fans out northwards and southwards (3). It returns to ground level when it is about one third of the way to the poles (4). This creates an area of high pressure at ground level (5). From here the air fans out again. Some moves back towards the equator as *trade winds* (6). The rest of the air continues towards the poles as south-westerly winds in the northern hemisphere or north-westerly winds in the southern hemisphere (7).

At the poles the air is very cold and sinks to ground level (8). Areas of high pressure are found at ground level (9). The cold air then flows away from the poles (10). When it

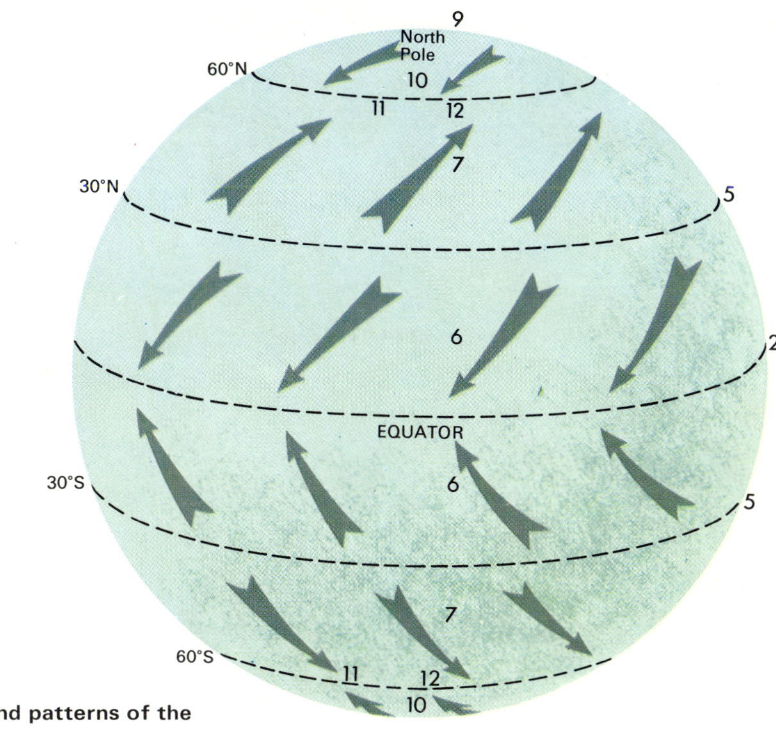

The wind patterns of the world

A section view of the air between the tropopause and the earth's surface in the Northern Hemisphere

is about a third of the way to the equator it meets the warm air coming from the tropics. The zone where they meet is called the *polar front* (11). The warm air rises above the cold air. At higher levels some of it flows back towards the equator (12).

4. Make an enlarged copy of the diagram below left. Find words and phrases in the text to replace the numbers.

Movements of the pressure belts

In general, the earth's pressure belts and wind systems always have the same pattern. But their exact positions above the earth's surface vary from season to season. In late December the sun is directly overhead at the Tropic of Capricorn. The low pressure belt is near to that Tropic. All the other pressure belts will be at their most southerly positions. In late June the sun is directly overhead at the Tropic of Cancer. The pressure belts will then be at their most northerly positions.

5. Make an enlarged copy of the diagram at the top of this page. Replace the numbers by the following words: equatorial low pressure, tropical high pressure, polar front, depression belt, polar high pressure, trade winds, south-west winds, north-west winds.

6. Which two times of the year are represented by your drawing of the position of the pressure belts in question 5? Explain your answer.

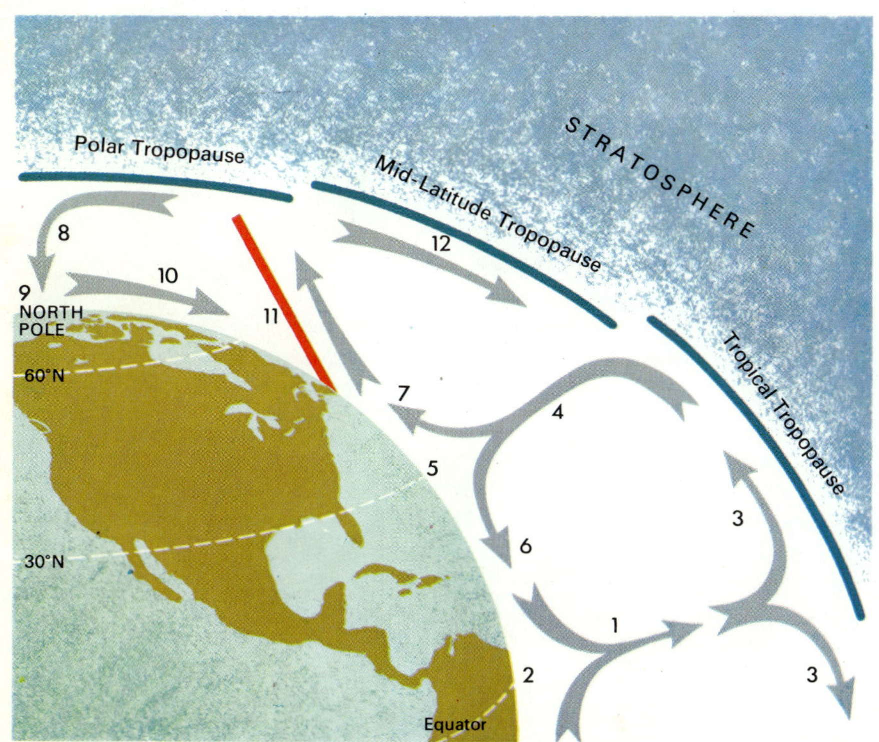

Measuring the Weather

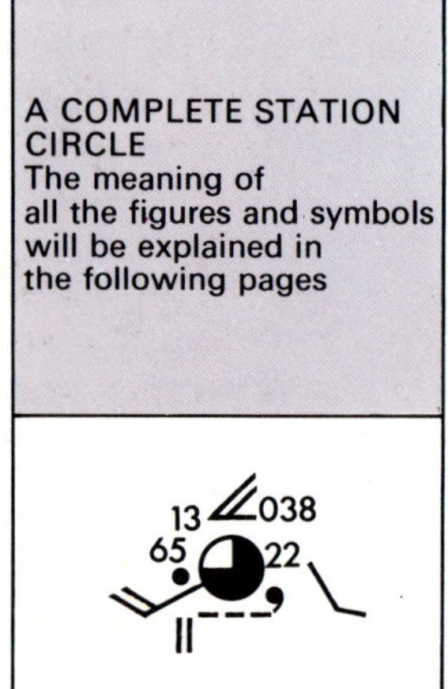

A typical weather recording station. All the instruments are kept inside an enclosed area.

Many of the observations made by weather stations are recorded on weather maps using the symbols shown in the complete station circle below.

A COMPLETE STATION CIRCLE
The meaning of all the figures and symbols will be explained in the following pages

Before any weather forecast can be made the state of the earth's atmosphere has to be carefully recorded. This is done at hundreds of weather stations around the world at fixed times called *synoptic hours*. After the measurements have been made the information is exchanged quickly. Maps and diagrams are drawn to show the weather which is being experienced at these times, including pressure, temperature and humidity. From these weather men can then forecast what may happen in the near future.

At a weather station all the instruments for measuring are kept in an enclosed space in an exposed position away from buildings and trees. The station is usually set out on grass and not on a concrete base. In the diagram above, the metal canister sunk into the ground is a *rain gauge*. The large box with slotted sides is a *thermometer screen*. Air can pass through this freely, but the direct rays of the sun cannot get inside. Various types of air thermometers are kept in it. Another thermometer is resting on two pegs just above ground level. This is a *grass minimum thermometer*. Other L-shaped thermometers are let into the ground to record the temperature of the soil. The glass sphere on the stand is a *sunshine recorder*. On the tall pole there is a *wind vane* for recording the direction of the wind and an *anemometer* for measuring its speed.

The next few pages describe how measurements are made by these instruments and one or two others. If your school has any of these instruments or if you have a simple weather station, you will be able to understand more easily how each measurement is made.

1. Listen to a weather forecast on the radio or television and list all the things that make up the weather.

Precipitation and the Rain Gauge

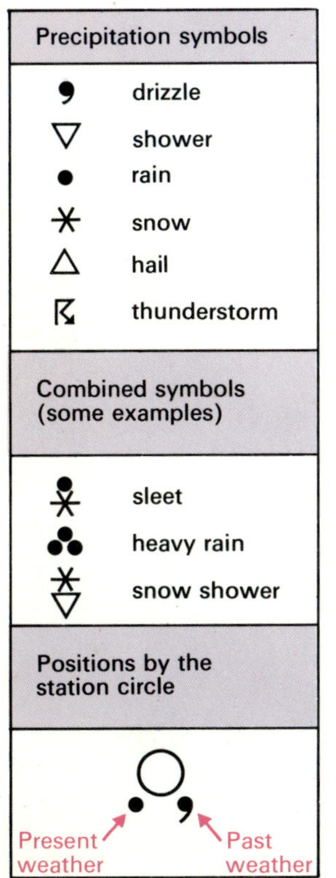

What is precipitation?

The word *precipitation* means all the types of moisture which are deposited on the earth's surface. Rain is by far the most important. Snow, sleet, hail, dew and frost are others.

Sometimes the water in the troposphere is invisible as water vapour. At other times the water can be seen as droplets in the form of clouds, mist or fog.

Rain, snow, sleet and hail fall from clouds to the earth's surface when certain conditions are found in the troposphere. *Cloud droplets* are very tiny drops of water. Hundreds of thousands of these are needed to make one rain drop heavy enough to fall to the ground. Only then will rain fall. Drizzle is made of smaller, lighter drops than rain.

Snowflakes are ice crystals which have been formed in a cloud where the temperature is below freezing point. The air temperature at ground level where they fall must also be below, or near, freezing point.

Sleet is a mixture of snow and rain. Hailstones are pellets of ice which have been made in some types of large clouds. When

Heavy winter snow blocking a mountain road

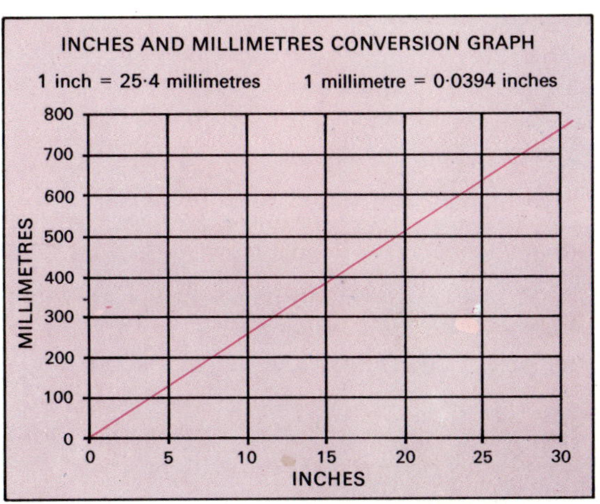

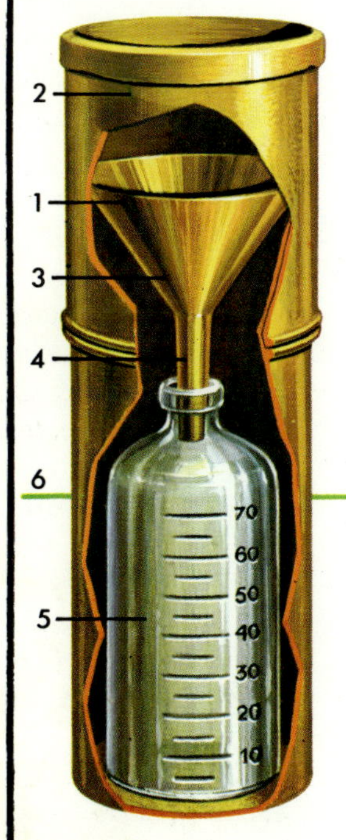

The rain gauge consists of a metal funnel (1) fitting into the top of a large metal canister (2). All forms of precipitation are collected in this and pass down the sloping sides of the funnel (3). Through the narrow copper tube (4) they pass into the smaller inner canister of a collecting bottle (5). Many gauges have a funnel 127 mm in diameter and are sunk into the ground so that the top of the funnel is 305 mm above the level of the grass (6). Each morning the funnel is lifted off the outer canister and the collecting bottle is lifted out carefully. The water in the bottle is then poured into a measuring cylinder (7).

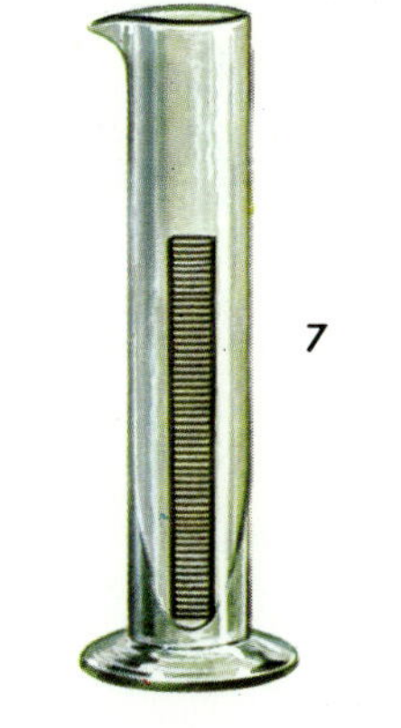

This has a much smaller diameter than the bottle. The depth of the water in the measuring cylinder is thus far greater than it was in the collecting bottle. It is therefore possible to record the daily precipitation with greater accuracy.

their weight can no longer be supported by the up-currents in the cloud they fall to the ground. Dew and frost are explained more fully on page 14.

Measuring precipitation

All forms of precipitation are collected in the inner container of a *rain gauge*, such as the one shown above. Snow, frost and hail must be melted first. A depth of 10 millimetres of snow is equal to 1 millimetre of rain. The total amount of water is then emptied into the measuring cylinder once a day at 09.00 hours. This total is recorded against the previous day's date.

Today precipitation is usually measured in millimetres. In this book millimetres are always used. In some countries records used to be kept in inches. To help you, a conversion graph is printed on this page.

1. Collect precipitation figures for a whole month at your school and make a graph like the one on page 32. Remember to show each day, even when there is no precipitation.

Temperature, Humidity and Thermometers

Temperature

Temperature is a measure of how hot or cold something is. It is measured by a scale divided into degrees. Weather men measure the temperature of the air. In the past temperature was measured in some countries by the *Fahrenheit* scale. Today weather men use the *Centigrade* scale. Centigrade is always used in this book. There is a conversion graph for degrees Fahrenheit and degrees Centigrade at the bottom of this page.

Humidity

Many New Yorkers go north for their summer holidays. They prefer the weather of New England to that of resorts further south. Of course, it is hotter in the south. But it is also much more uncomfortable because the air is so *humid*.

Air contains a small amount of water vapour. The amount changes from time to time and from place to place. It is important in giving the 'feel' to the air. Is it muggy or scorching, keen or raw? The weather man has a more exact method of describing the amount of water vapour present in the air.

Air is said to be *saturated* when it can hold no more water vapour. The amount of water vapour which it can hold depends upon its temperature and pressure. Cold air holds less water vapour than warm air before it becomes saturated. When air is saturated

A hot, dry summer day in Granada, Spain

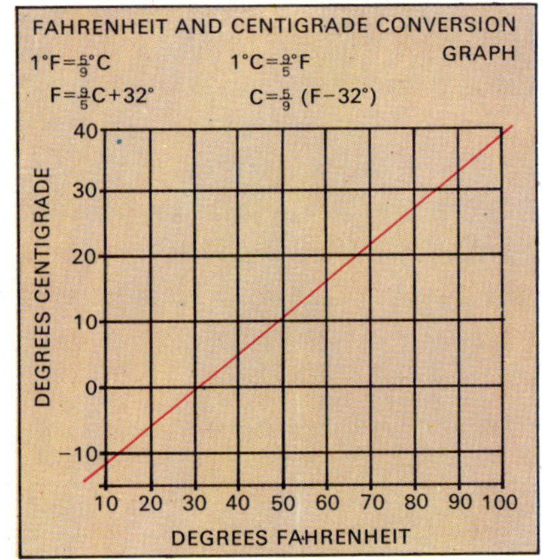

the water vapour turns into droplets of water. These form mist, fog or a cloud.

The weather man compares the amount of water which is present in the air with the amount that there would be if the air was saturated. This is called *relative humidity* (RH) and he works it out as a percentage. RH 96% might be muggy or raw air. RH 47% might be scorching or keen air.

1. What word would you use to describe the weather in the photograph on this page: muggy, scorching, keen or raw?

Thermometers

Dry bulb thermometers. The temperature of the air is recorded by a *dry bulb thermometer*. This is a narrow, sealed tube of glass with a bulb at one end. Inside it is mercury, which expands and contracts with changes in the air temperature. The end of the mercury column shows the air temperature.

Weather men also record the highest and lowest temperatures every 24 hours. This is done by using two special thermometers. The *maximum thermometer* records the highest temperature. It is like a dry bulb thermometer, except that the cavity in the glass has a small constriction near the bulb. As the temperature rises, the mercury expands. As the temperature falls, the constriction prevents the mercury from returning to the bulb. In this way the highest temperature during any period of time is shown by the end of the mercury further from the bulb.

The *minimum thermometer* measures the lowest temperature. It is filled with alcohol instead of mercury. It has a pin inside the alcohol. The pin is carried toward the bulb as the temperature falls and the liquid contracts. The alcohol expands when the temperature rises again, and the pin is left in the tube with the end further from the bulb marking the minimum temperature.

Both the maximum and minimum thermometers are mounted horizontally to prevent gravity from affecting the readings. Both need to be reset each day. The maximum thermometer is shaken until the mercury has returned to the bulb. The other is tilted away from the bulb until the pin is just touching the top of the alcohol.

2. Suppose that one morning at 09.00 hours you go to the thermometer screen to take the temperature readings. The air temperature is 15°C. The maximum temperature on the previous day was 25°C. The minimum temperature during the previous night was 12°C. Draw pictures of the thermometer to show:
(a) what they look like as you take the readings;
(b) what they will look like after resetting.
3. Keep records of the dry bulb, maximum and minimum temperatures at your school over a period of time and graph them.

Wet bulb thermometers. The humidity of the air is measured with the aid of a *wet bulb thermometer*. This thermometer is set up with its bulb covered by a small piece of muslin. This is connected by a thin wick to a con-tainer of distilled water. By this method the bulb is surrounded with saturated air. Its temperature is never higher than that of the dry bulb thermometer kept next to it because the air surrounding the dry bulb thermometer can never be more than saturated.

To calculate the relative humidity of the air both dry and wet bulb readings are taken. Weather men use special tables of figures to find the relative humidity. Dry and wet bulb temperatures are written on weather maps, but relative humidity is not.

4. When you run a hot bath in a closed room the air soon becomes saturated. Mist or a cloud is formed. If you then open the window and let in some colder air the mist will become thicker at first. It may clear later. Explain these events.
5. Keep records of the wet bulb temperatures at your school. Obtain a set of relative humidity tables. Work out the relative humidity for each day.

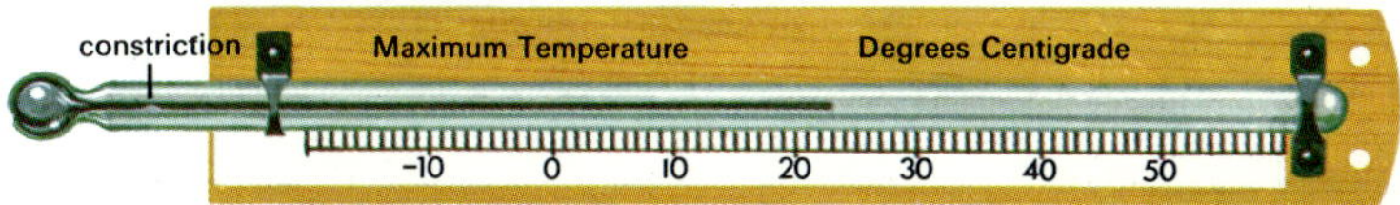

MAXIMUM THERMOMETER

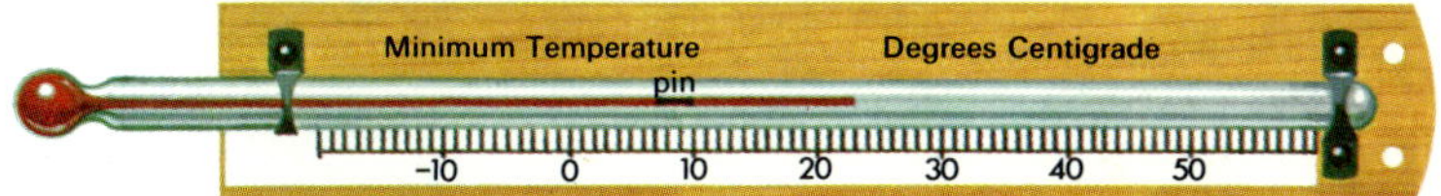

MINIMUM THERMOMETER

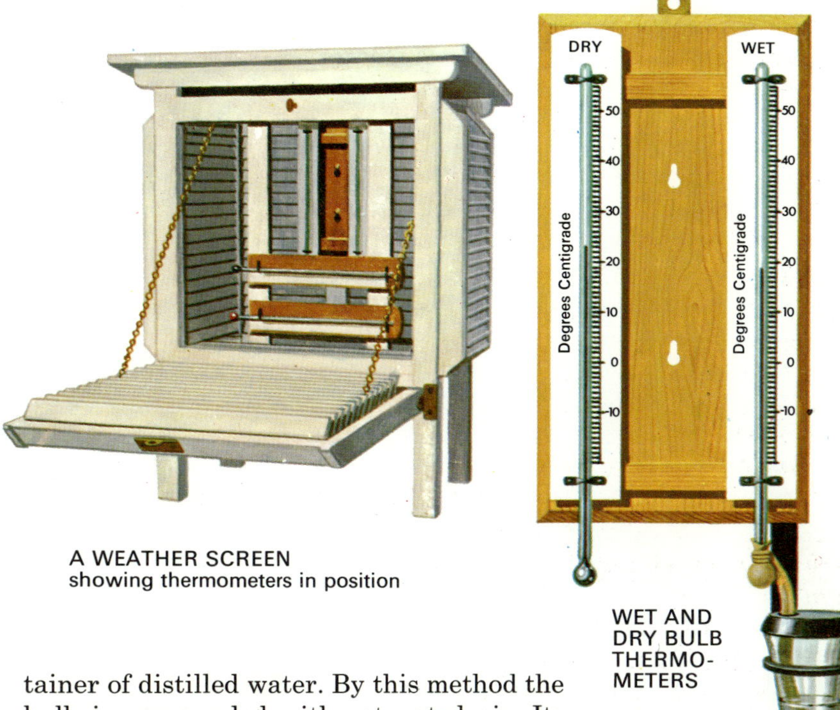

A WEATHER SCREEN
showing thermometers in position

WET AND DRY BULB THERMO-METERS

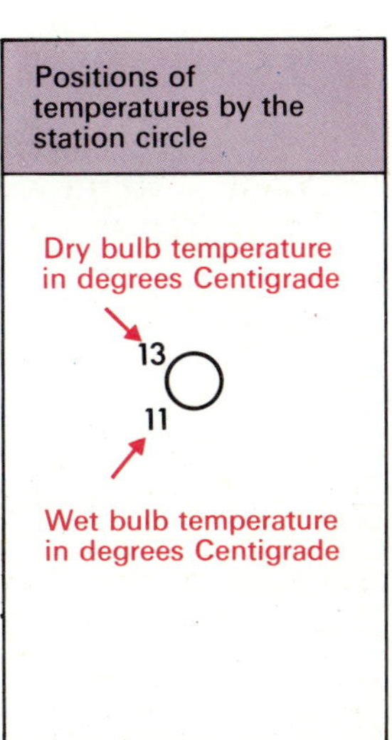

Sunshine and Cloudiness

Sunshine

Seaside resorts try to attract visitors by claiming that they have a better sunshine record than rival towns. They base their claims on the total hours of sunshine registered each day by an instrument called a *sunshine recorder*.

The most widely used recorder, the Campbell-Stokes, is shown below. Each day a specially designed card is slotted into the semicircular frame of the instrument. When the sun is shining the large glass ball mounted in front of the card focuses the sun's rays onto it. In this way a mark is burned on the card at those moments when the sun shines. As the earth spins, the position of the burn mark moves along the card. No burn marks are made when the sun is not shining. The card is marked in hours and parts of hours. It is therefore easy to add up the length of time during which the sun shone on any day.

A Campbell-Stokes sunshine recorder. Below is the record card of sunshine for a September day.

A summer day with five-eighths of the sky covered with cloud. The clouds are being seen at an angle here. It is therefore not possible to see the estimated three-eighths sky which exists between the clouds.

1. The picture at the foot of this page shows a sunshine record for a September day. For how long did the sun shine?

Cloudiness

Sometimes the sun is hidden by clouds or by fog or mist (see page 15). The weather man is interested in how much cloud there is in the sky and also what type of cloud it is (see pages 16 and 17).

The amount of cloud is measured in *oktas*, or eighths of the sky. It is sometimes difficult to estimate this figure. Many clouds are seen from below at an angle, and gaps between them cannot be seen. By first looking at the sky overhead it becomes easier to estimate cloudiness.

As with air temperature and pressure, records of cloud amounts are made at an exact moment in time. Cloudiness can therefore be recorded on each station circle. The centre of the circle is used for this, following the system shown in the diagram on the right.

2. Look back to the map on page 1. Which stations reported no cloud, where was it completely overcast (8/8ths cloud) and where was the sky obscured by mist or fog?

Symbols for cloud cover	
◯	clear sky
◔	one okta
◔	two oktas
◑	three oktas
◑	four oktas
◕	five oktas
◕	six oktas
◖	seven oktas
●	eight oktas
⊗	sky obscured

These symbols are always positioned inside the station circle

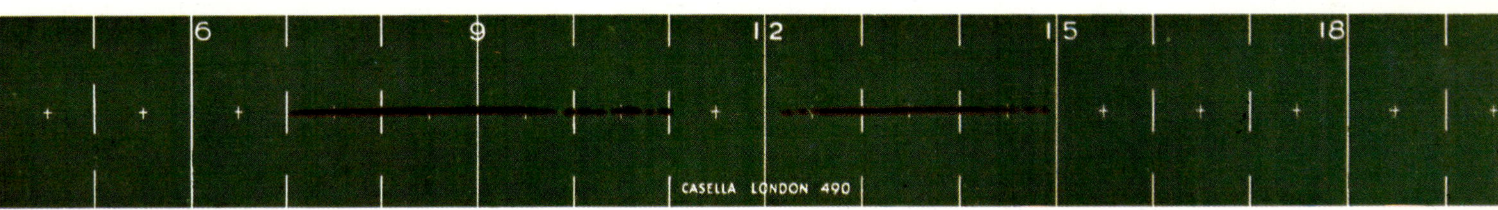

Pressure and Barometers

A mercury barometer

An aneroid barometer

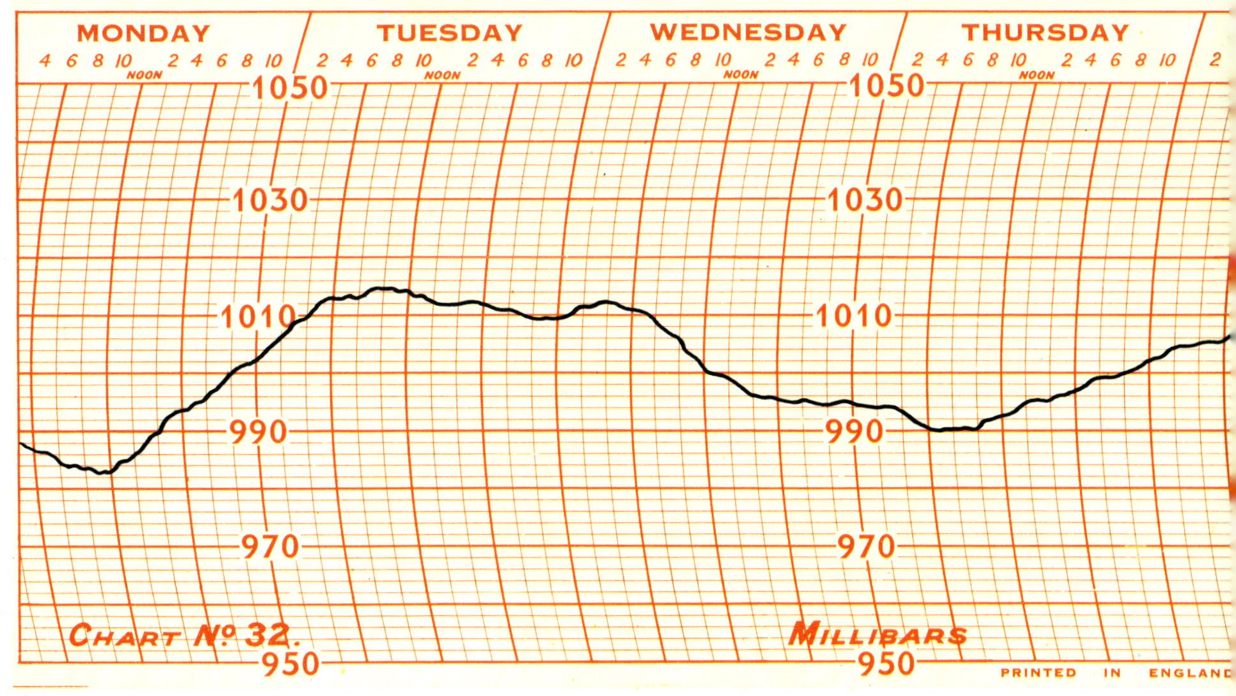

Pressure

Pressure is the weight of the air at any point on the earth's surface. The higher you go into the troposphere, the lower the pressure becomes. It is for this reason that modern aircraft which fly at great heights need to have *pressurised* cabins. Pressure also varies slightly from time to time at the same place on the earth's surface.

1. Look at the diagram on page 4. What is the pressure reading in millibars—

(a) where the Concorde aircraft is flying?

(b) 100 kilometres above the earth?

Barometers

There are two main ways of measuring pressure. The more accurate is by a *mercury barometer*. It consists of a column of mercury in a glass tube. It is held there by the weight of the air pressing on a small diaphragm at its base. The tube has a scale beside it.

The other way to measure pressure is by an *aneroid barometer*. Many homes have one hanging on the wall. It consists of a small metal box out of which some of the air has been taken. As the outside pressure changes the box expands and contracts. A system of levers coupled to a pointer records the pressure on a dial. Sometimes these levers stick a little. If you tap the glass the pointer will move and show the correct reading.

Weather men need a continuous record of pressure. To make this, an aneroid barometer is coupled to a lever with a pen at its end. The pen presses against a drum which makes one complete turn each week. A sheet of graduated paper is fitted to this drum each Monday morning. During the week the pen traces a continuous record of the pressure. This instrument is called a *barograph* and the record sheet a *barogram*.

Part of a barogram is shown above. The pressure has been recorded on it in *millibars*, which are the international unit for measuring pressure.

Weather maps show lines which are drawn through all the places that have the same pressure. These lines are called *isobars*. Air pressure figures are always adjusted to the readings that would be recorded at sea level. In this way pressure differences caused by height are not mapped.

2. Aneroid barometers often have the words 'Stormy', 'Rain', 'Change' 'Fair', and 'Very Dry' on their faces. How accurate do you think these words are?

Symbols for pressure and pressure trends (some examples)		Positions by the station circle
—	steady	Actual pressure in tenths of millibars. The initial 9 or 10 are omitted
\\	falling	038
/	rising then steady	22
\\/	steady then falling	Change of pressure in the last 3 hours in tenths of millibars
/\\	rising then falling	Pressure trend

Wind Direction and Wind Speed

Wind affects our lives in many ways. It may be useful or a nuisance. In the past wind was used to drive large sailing ships around the oceans. Today sailing ships are more often small pleasure boats, like the one shown below.

1. Make a list of the ways in which you think wind has been useful to man. Then list ways in which you think it is a hindrance to him.

Wind direction

On top of many old country churches and public buildings decorative iron *wind vanes* or *weather cocks* can still be seen. They show the direction of the wind. Weather men use a wind vane to record wind direction.

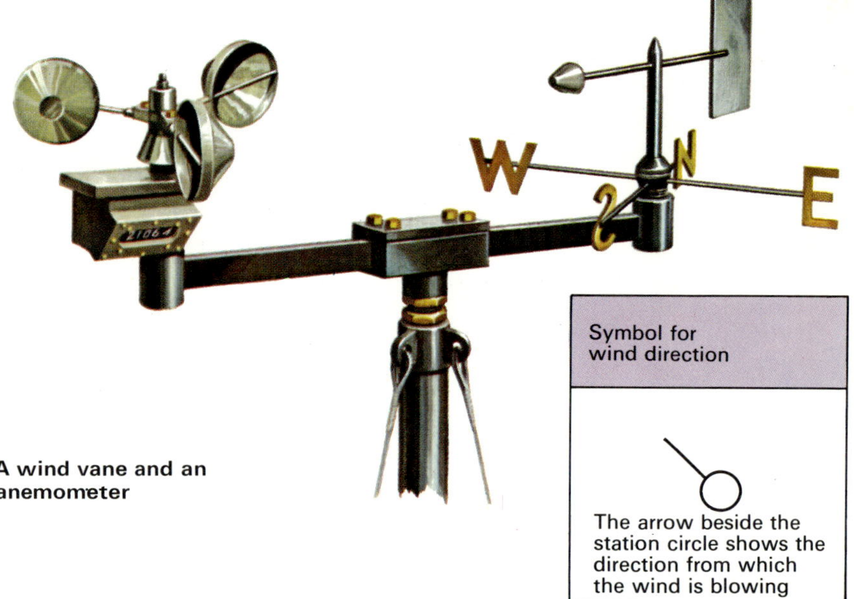

A wind vane and an anemometer

Symbol for wind direction

The arrow beside the station circle shows the direction from which the wind is blowing

A yacht in a fresh breeze

The vane is simply an arrow which turns on a pivot. Its tail is larger than its head. When the wind strikes the larger area of the tail it turns the arrow to point in the direction from which the air is coming. This is used as the name of the wind. On the weather map an arrow is drawn on the side of the station circle from which the wind is blowing.

Wind direction can be described in two ways. One way is to name the compass point from which the wind is blowing, for example, north, west, south-east. The other way is to give the compass bearing, such as 000°, 270°, 135°. By this method, the compass face is divided into 360°. North is called 000°, east 090°, south 180° and west 270°. Other compass points have numbers which come between these.

2. Give the compass point names for each of the following winds: 045°, 112½°, 315°, 292½°.

3. Give the compass bearings for each of the following winds: NNE, NNW, ENE, WSW.

Air generally moves from areas of higher pressure to those of lower pressure. If the earth did not rotate, then winds would blow straight from the centres of high pressure areas to the centres of neighbouring low pressure areas. However, the rotation of the earth causes this movement to be deflected.

If you look at a map showing both isobars and wind direction it will be seen that the air

THE BEAUFORT SCALE

Beaufort Number	Description of the Wind	Speed of the Wind in Knots	Possible Effects of the Wind
0	CALM	Less than 1	Smoke rises vertically
1	LIGHT AIR	1–3	Direction of the wind is shown by smoke drift but not by the wind vane
2	LIGHT BREEZE	4–6	Wind felt on the face; leaves rustle; wind vanes are moved by the wind
3	GENTLE BREEZE	7–10	Leaves and twigs are in constant motion; the wind extends a light flag
4	MODERATE BREEZE	11–16	Dust and loose paper are raised; the small branches of trees are moved
5	FRESH BREEZE	17–21	Small trees begin to sway; crested wavelets form on inland water
6	STRONG BREEZE	22–27	Large branches of trees begin to move; telegraph wires whistle; umbrellas are difficult to use
7	MODERATE GALE	28–33	Whole trees are set in motion; some difficulty in walking against the wind
8	FRESH GALE	34–40	Twigs break off trees; great difficulty in walking against the wind
9	STRONG GALE	41–47	Slight structural damage is caused; chimney pots and slates are removed
10	WHOLE GALE	48–55	Trees are uprooted; much damage to buildings is caused. This wind is seldom experienced inland
11	STORM	56–63	Widespread damage occurs. This wind is rarely experienced in Britain
12	HURRICANE	More than 64	The whole countryside is devastated

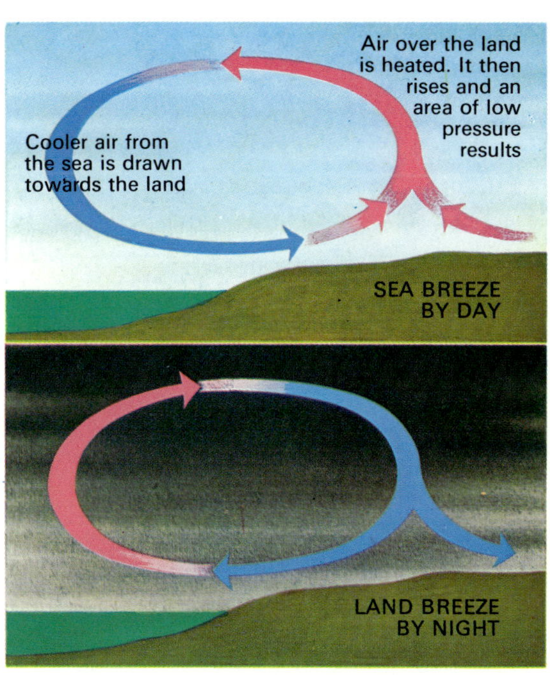

How land and sea breezes are caused

Symbols for wind speeds

◎	calm
	1-2 knots
	3-7 knots
	8-12 knots
	13-17 knots
	18-22 knots

Further half feathers are added up to 4½, which show a wind speed of 45 knots. Then—

	50 knots or more

generally moves parallel to the isobars. In the Northern hemisphere, if you stand with your back to the wind the lower pressure will be on your left-hand side. In the southern hemisphere, if you stand with your back to the wind the lower pressure will be on your right-hand side. This is called *Buys Ballot's Law.*

4. Look back to the map on page 1. Does Buys Ballot's Law hold true there? From which direction would the wind be blowing if you were in (a) Cardiff (b) Glasgow (c) Copenhagen (d) the Shetland Isles?

Wind speed

The closer together the isobars, the stronger will be the wind. Wind speed is recorded by an *anemometer.* One type has a set of three cups. Each is at the end of an arm joined to a central pivot. The anemometer turns in the moving air and is coupled to a device for measuring the wind speed. This may be a dial like a car speedometer. It may be a box which produces rapid sounds whose rate varies with the wind.

Wind speeds are plotted on weather maps by putting tails onto the wind direction arrow. Each full tail means a speed of 10 knots (nautical miles per hour); each half tail means 5 knots. When there is no wind at all a larger circle is drawn round the station circle.

An alternative method of determining the wind speed is to use the *Beaufort Scale,* shown on the left above.

5. Look back to the map on page 1. Where are the winds strongest and what are their speeds? Are these the places where the isobars are closest together? Where is there a calm?

6. How useful is the Beaufort Scale (a) in towns (b) in the country (c) at sea?

Land and sea breezes

When you are lying sunbathing on a beach you may have wondered why a breeze is often blowing from the sea. At the coast, air may move from sea to land during the day. At night it may move in the opposite direction. These movements over a small area are made in a similar way to the movements of the world's wind belts on a much larger scale.

Land gains and loses heat more quickly than sea. On a summer's day the air above the land is slightly warmer than that over the sea. It rises, and an area of slightly lower pressure is made. Air from the sea moves *on-shore* to replace the air that has risen.

At night the opposite happens. The land cools down more quickly than the sea. The air over the sea is then slightly warmer than the air over the land. An area of slightly lower pressure is made over the sea. Air then moves *off-shore* from the land.

7. The two diagrams on this page show the air movements in land and sea breezes. Make labelled copies to explain what is happening.

Dew and Frost

The ground radiates heat at certain times (see page 4). At night this process slows down and may stop completely. Then the air near the ground may be cooled rapidly, especially when the air is still and there is little wind. In such conditions dew and frost may form.

Dew

If you go out into the garden or onto the school playing-field early in the morning, you may get your shoes wet, even though no rain fell during the night. This is *dew*.

Air which is moist does not have to be cooled much before it becomes saturated (see page 8). If the air is cooled further, then it can no longer hold the water vapour. Instead, the vapour changes into small water droplets. The temperature at which this happens is called the *dew point*. The dew is usually deposited on cold surfaces, such as the tops of cars or on blades of grass.

At times moisture which has been deposited as dew will be cooled still further until it freezes. This is frozen dew. It is different from frost.

Early morning dew drops glistening on a spider's web

On a cold winter morning hoar frost covers the ground and rime frost coats the trees and telegraph poles.

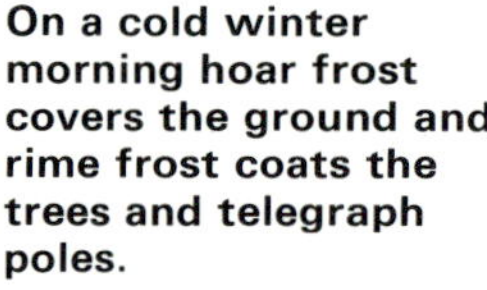

Frost

Most people have seen, early on a winter morning, the fern-like patterns made by frost on window panes. Frost, like dew, is made from water vapour. But dew is formed when the air temperature is above freezing point. If the air temperature is below 0°C when saturation occurs, the moisture in the air will be deposited as crystals of ice. On windows or the ground this is called *hoar frost*.

Sometimes water droplets can be found in air which has a temperature below freezing point. These droplets are said to be *super-cooled*. When super-cooled water droplets touch any object with a temperature below freezing point they turn into ice. In this way, trees, telegraph poles and gate posts are coated with a layer of *rime frost*.

Glazed frost, or *black ice*, is a third form of frozen water at the earth's surface. If rain falls onto ground which has a temperature below freezing point, the rain freezes and becomes a sheet of ice. This is the danger to motorists sometimes mentioned on the radio.

1. List the ways in which frost can be a hindrance to man. Can you think of any ways in which it helps him?

2. If you stand at a bus stop on a cold winter's night a thin covering of ice may form on your coat. Explain what is happening.

Mist and Fog and Visibility

Visibility

At each synoptic hour weather stations record the distance of the furthest object which can be seen. Each station has a list of physical features, buildings, trees and other objects which are at known distances from the station. This is the distance of *visibility*.

1. Look back to the picture on page 6. List the objects which could be used to find the visibility.

When visibility is less than 1 kilometre, *fog* is recorded. If the visibility is between 1 and 2 kilometres *mist* is recorded. Visibility is noted and plotted in tenths of kilometres. The station circle on this page shows how this is done.

Fog is like a cloud at ground level. It is created by the cooling of moist air and the condensation of water vapour into droplets. It can be formed in a number of ways. The two most common types are *radiation fog* and *advection fog*.

Radiation fog

In autumn and winter many nights are calm and clear. The air near the ground cools rapidly. If it reaches its dew point, droplets of water collect in the air and form a fog. This type of fog is more common in open countryside than in towns. It also collects in valley bottoms. It sinks to the lowest level possible because cold air is heavier than warm air.

Advection fog at the Golden Gate Bridge, San Francisco

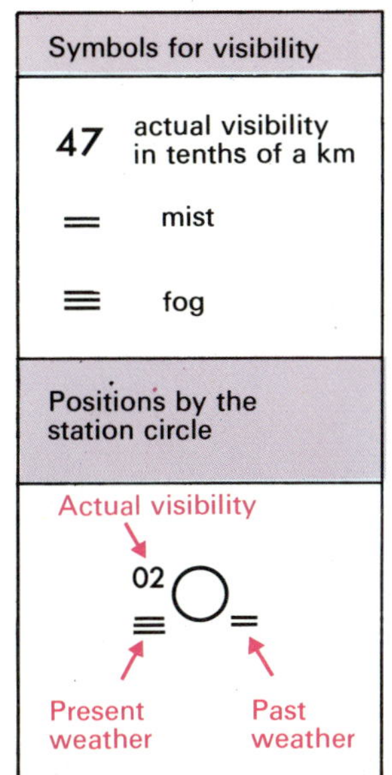

Advection fog

When moist air passes across a warm surface to a colder surface it is cooled from below. In this way it may have its temperature reduced to the dew point and a fog will result. This often happens along a coast in spring, when air moves from the warmer sea surface to the colder land surface. In this case it is called *sea mist* or *coastal fog*.

The photograph above shows coastal fog enveloping the Golden Gate Bridge at San Francisco. Here, moist air moving slowly towards the land is cooled by a cold current just off the coast. Fog forms and then drifts inland.

2. Copy the diagrams below. Add labels to each to explain how radiation fog and sea mist are formed.

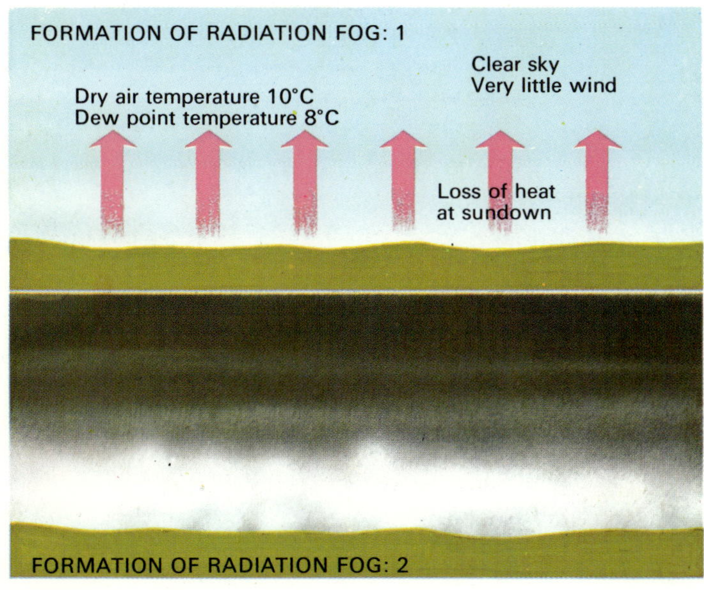

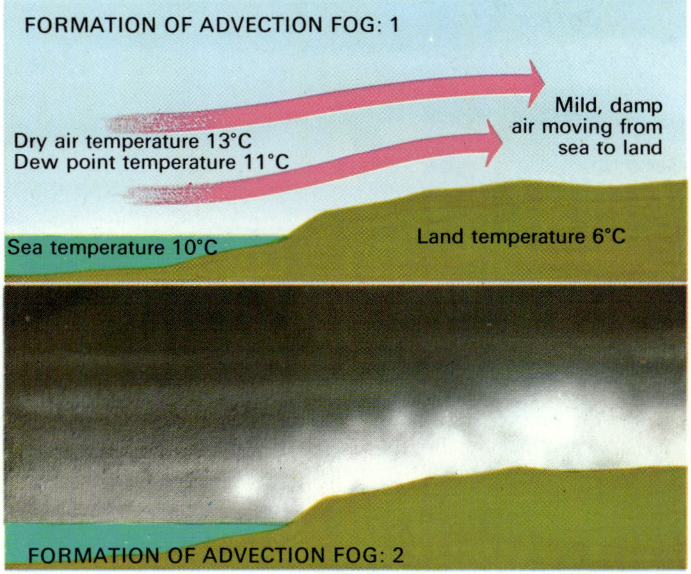

Cloud Types

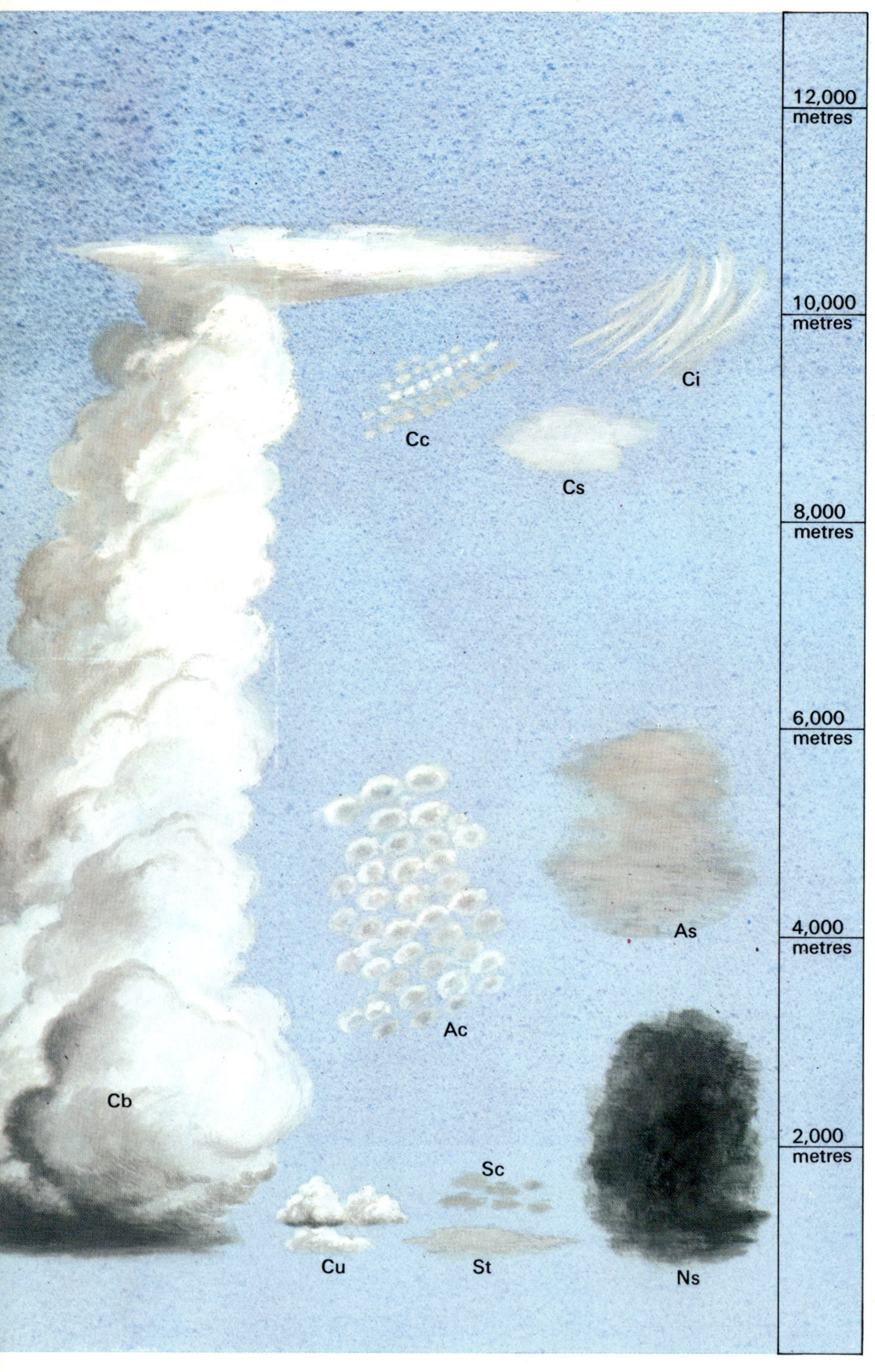

The ten main types of cloud, showing their positions in the troposphere and their general shapes

Clouds greatly influence our weather. They are all found within the troposphere to a height of about 16 kilometres. There are three main bands of clouds. High-level clouds have names which include the word *cirro* or *cirrus*. Medium-level clouds have names starting with the word *alto*. Low-level clouds have various names.

Some clouds also have a common name that describes their appearance. Clouds which occur in great sheets are given a name including the word *stratus* or *strato*. These words mean 'a layer'. Clouds which occur in a heaped form are given a name which includes the word *cumulus* or *cumulo*. These words mean 'a pile'.

HIGH-LEVEL CLOUDS (above 6,000 metres)

These usually appear as a thin, feathery, white veil that does not completely blot out the sun. They are made of ice crystals, not cloud droplets, because the air temperature in the upper troposphere is usually below freezing point. There are three types:

Cirrus (Ci). The ice crystals are drawn out into feathery streaks across the sky. Cirrus clouds are sometimes called *mare's tails*. See the picture on page 22.

Cirrostratus (Cs). When there are a lot of ice crystals at high levels, cirrostratus clouds form. They make a complete layer of white cloud covering large parts of the sky. If the sun or moon shines through, it will have a halo.

Cirrocumulus (Cc). If the ice crystals are gathered into small clouds arranged in rows, the name cirrocumulus is used.

MEDIUM-LEVEL CLOUDS (2,000 to 6,000 metres)

There are two main types of clouds at medium levels. They are similar to the higher clouds but look heavier and darker.

Altostratus (As) is a thin layer of cloud droplets or ice crystals which allow the sun or moon to shine through weakly. It can develop to a thickness of hundreds of metres. It will then blot out the sun or the moon.

Altocumulus (Ac) are small cumulus or piled-up clouds at medium heights. These clouds are often arranged in rows. The amount of blue sky which can be seen between the clouds depends on their size and the spacing between them. Such a cloud cover may be called a *mackerel sky*. See the pictures on the opposite page and on page 2.

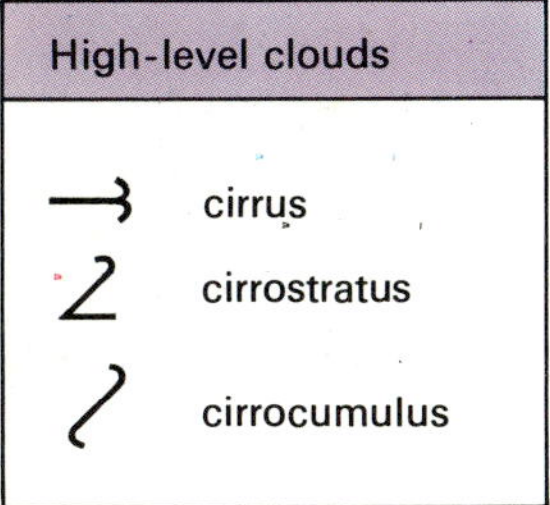

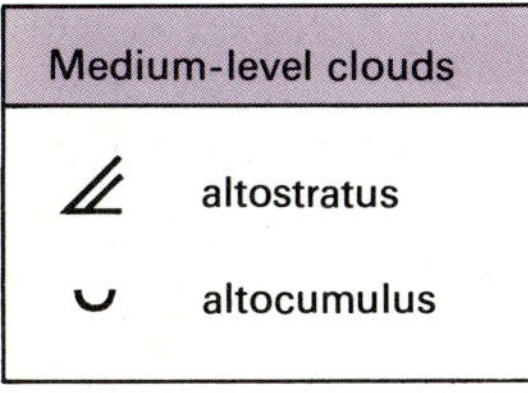

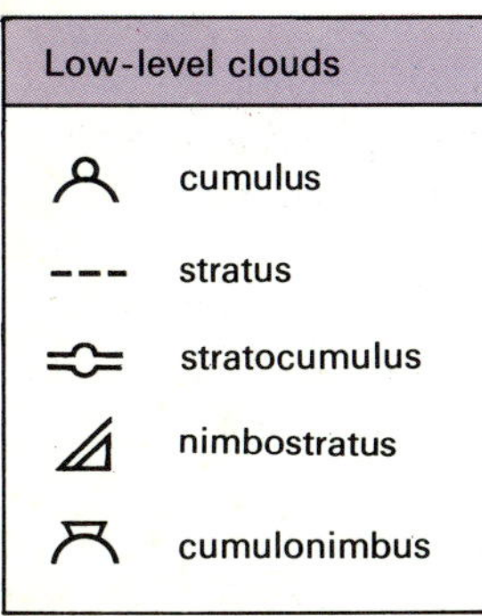

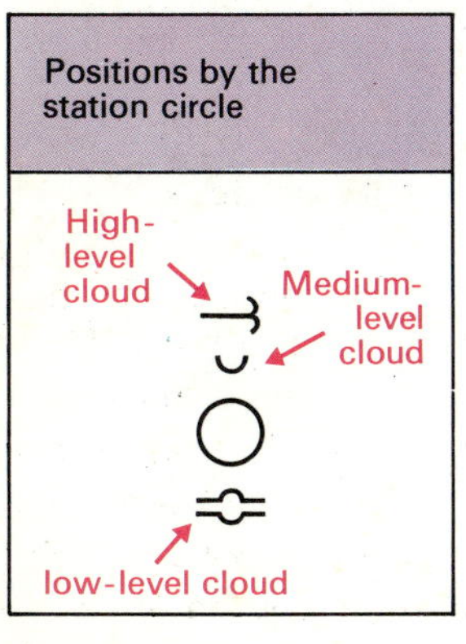

LOW-LEVEL CLOUDS
(mainly below 2,000 metres)

There are five main types of cloud which have their bases below 2,000 metres:

Cumulus (Cu) are thick clouds with a flat base and clearly outlined domes or billows above. Small clouds with a lot of clear sky between them are called *fair-weather cumulus*. Much larger clouds are called *towering cumulus*, and are likely to bring rain.

Stratus (St) is a simple layer of cloud which has its base between ground level and about 2,000 metres. It has no clear shape, has a depressing, grey appearance and may be many hundreds of metres thick. Stratus cloud may bring periods of drizzle or rain.

Stratocumulus (Sc) is a much lower and much larger form of cloud than altocumulus. It is a cloud layer consisting of a large number of cumulus clouds with the same base level. The upper parts of the clouds are blurred. Often individual clouds appear in bands or rolls. Between these the sun may shine. The amount of sun depends on the size of the breaks between the clouds.

Nimbostratus (Ns) forms the largest layers of cloud. It can be part of a continuous layer of cloud from only a few metres above the ground right up to the levels of medium and high-level clouds. It produces a large proportion of the rain that falls in temperate climates. When looking at it from the ground, you can distinguish nimbostratus from ordinary stratus by looking for small masses of broken cloud beneath its solid base. These are called *scud*.

Cumulonimbus (Cb). When cumulus cloud towers up to its greatest heights it eventually approaches the *tropopause*, the upper limit of the troposphere. Here the upward currents weaken. Strong winds blow across them and carry the ice crystals in the cloud horizontally. In this way, the so-called 'anvil' of the cumulonimbus cloud is created. This is a thunder cloud, and it can produce very heavy rain or hail as well as thunder and lightning.

1. The diagram on the opposite page shows all the positions and forms of the clouds. On the right there are also some photographs of clouds. Make a sketch of each, and identify the clouds.

2. Look at the clouds in the sky today. Make sketches and give the names of those which are not shown in the photographs here.

Altocumulus cloud at a height of 2,500 metres seen from an aircraft flying at a height of 7,000 metres

Above and below: the top and base of a bank of cloud. The base is at 600 metres, the top at 4,000 metres.

Air Masses and Air Streams

You must remember three important points when studying air masses and air streams:
1. The planetary wind pattern (see page 5).
2. Buys Ballot's Law, giving the relationship between pressure patterns and wind direction (see page 12).
3. The relationships between the closeness of isobars and wind speeds (see page 13).

A table of air streams affecting the weather of north-west Europe

Air masses

In some parts of the world air remains relatively still for long periods of time. These are areas of high pressure. In these areas large volumes of air can develop similar patterns of temperature and humidity. These volumes of air are then called *air masses*. The places where they develop are called *source regions*. When a mass of air later moves away from its source region to another part of the world it keeps its temperature and humidity for a few days. Source regions influencing European weather are listed below.

NAME OF AIR STREAM	SOURCE AREA	DIRECTION OF MOVEMENT	POSSIBLE CHARACTERISTICS OF THE AIR STREAM AND ACCOMPANYING WEATHER	
MARITIME TROPICAL (mT)	Mid-Atlantic; Azores area	From the south-west	SUMMER	Warm and humid air. Fog on coasts, warmer inland when cloud breaks. Sometimes produces thundery conditions in the east.
			WINTER	Mild moist air. Overcast and muggy conditions.
CONTINENTAL TROPICAL (cT)	North Africa and the Mediterranean Basin	From the south and south-east	SUMMER	Not very frequent, bringing heat waves with long sunny periods. Some chance of thunder in Southern England
			WINTER	Very rare, bringing a warm spell with dry air and little cloud
MARITIME POLAR (mP)	Greenland, North Atlantic and northern North America	From the west or north-west	SUMMER	Cool air with variable humidity. Can be rather cloudy, especially in the west, bringing showers or heavier rain. At other times, crisp, clear air
			WINTER	Warmer than might be thought, because it has travelled over the warm water of the Atlantic. Showers and bright periods
CONTINENTAL POLAR (cP)	Siberia, Russia and the Baltic area	From the east or south-east	SUMMER	Dry and fairly warm, coming from the continental interior. Misty on the east coast, clear weather elsewhere
			WINTER	Brings the harshest weather with temperatures around freezing. Variable humidity depending on length of North Sea crossing. Variable cloud, with the heaviest falls of sleet and snow on the east coast
MARITIME ARCTIC (mA)	North Polar ice	From the north	SUMMER	Rare air stream from the north, bringing an unseasonal cold spell with some rain
			WINTER	Very cold, unstable air, bringing snow after its crossing of the Norwegian Sea. Snow on exposed east and north-east coasts and on high ground inland

Air streams

When an air mass moves away from its source region it is called an *air stream*. The air stream's temperature and humidity gradually change. Eventually the characteristics of the original air mass will be completely changed. But, for a period of some days the air stream will keep the characteristics which it had in the source region.

The British Isles are situated in a part of the globe which can be influenced by air from several source regions. The most important of these are shown in the table on the opposite page. This table gives the direction from which these air streams reach the British Isles. It also shows some of the more usual weather features associated with each. These weather features vary with the season of the year.

The name of an air stream tells you where it began as an air mass. If the air was originally stationary over a land area the air stream is called *continental*. If the air was stationary over an ocean the air stream is called *maritime*. The words *tropical*, *polar* and *arctic* are then added to show the latitude where the air stream began.

1. On the right are four maps showing isobar patterns over north-west Europe. All other weather information has been left out. Each map has a month printed beside it. For each one work out the following:

(a) the general direction of the winds;
(b) the probable name of the air stream;
(c) the source region of the air stream;
(d) the weather characteristics which might be found over the land areas.

Then write sentences to describe the weather on each map. For example, after studying the the first map you might write: 'North-west Europe is being influenced by easterly winds which are bringing a continental polar air stream from Siberia. Misty conditions may be found along the east coast of Britain but elsewhere it should be mainly clear and dry'.

It is important to remember that weather forecasting cannot be done only by looking at air streams and their characteristics. Many other things have to be recorded and interpreted when forecasting. Today weather men give less importance to air streams than they used to. But air streams are still one useful guide in forecasting the weather which lies ahead.

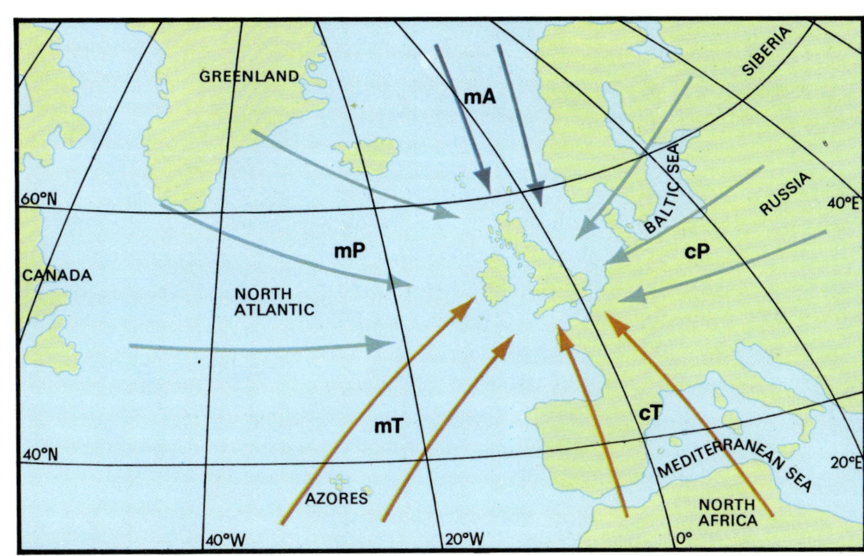

▲
Some routes followed by air streams affecting the weather of the British Isles

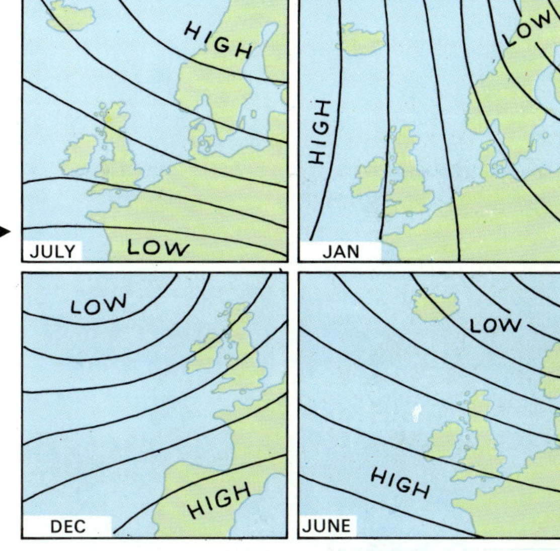

Four pressure maps of north-west Europe (see ▶ question 1)

Small cumulus clouds over central Norway. They have developed in a maritime polar air stream. The weather conditions which this air stream brings are described in the table on the opposite page.
▼

Warm and Cold Fronts

On page 5 it was clear that air streams meet one another at many places at the earth's surface and high above it. When this happens there will be a slow mixing of the two air streams. There is a zone in which the characteristics of the air alter rapidly from those of one air stream to those of the other. This zone is called a *front*.

The vertical arrangement of two streams of air can be shown by a diagram. The air stream which is colder and denser will remain at ground level. The warmer lighter air will ride over the top of it. The angle between the ground and the front has been exaggerated in the diagram on this page. The gradient of a front is rarely more than 1:30 and often as little as 1:300. These gradients are hard to draw on a small piece of paper.

The weather man can often identify the positions of fronts at ground level when enough stations have reported temperatures, pressures, cloud types and humidity. Once identified, the fronts are marked on the weather map, as on the map on page 1. If colder air is being displaced by warmer air it is called a *warm front*. If warmer air is

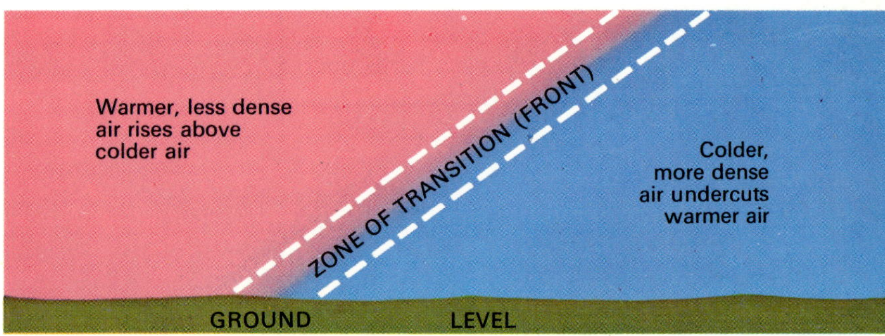

A day with a slow moving, highly active warm front. Prolonged heavy rain has saturated the ground. Low nimbostratus clouds are overhead, their base at about 150 metres. The calm sea shows that there is little wind.

Heavy rain as a front passes

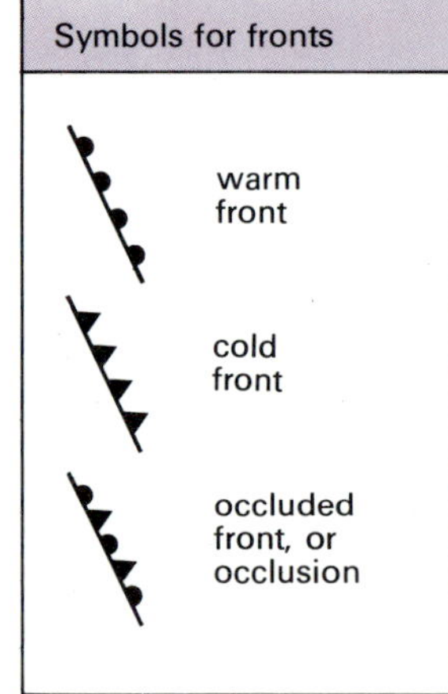

being replaced by colder air, then it is called a *cold front*. The standard symbols for these are shown in the diagram on the left.

Since the front is sloping so slightly it follows that its position at ground level may be hundreds of kilometres away from its position at higher levels. In both warm and cold fronts, warmer air is rising slowly over colder air. Cooling takes place and at various levels condensation occurs. Fronts are zones in which clouds of various types will be seen. They often bring long periods of rain.

1. When the weather man forecasts that a front will pass over the area in which you live, watch how the clouds change during the day. Relate the changes to all that has been said on pages 16 to 19.

2. What name would you give to the front shown in the diagram above if it was moving from (a) left to right (b) right to left?

Pressure Patterns

A complete weather map, such as the one on page 1, shows details of many of the things that make up the weather. The curved lines are the most noticeable. They seem to be drawn on the chart in a random fashion. They are very important in weather forecasting.

Each of these lines is called an *isobar*. It is like a contour on a map showing land heights. Just as contours join places of equal height above sea level, so isobars join places which have the same air pressure at any moment in time. When all the isobars have been drawn it is possible to see which areas on the map have low pressure and which have high pressure. Each pressure system often has its own pattern of weather.

Isobar patterns are usually made up of smoothly curved lines. The isobars only change direction noticeably where they cross a front. At these points they may be drawn with a sharp angle. Weather men therefore plot the fronts on the weather map before isobars.

1. Below is a map showing pressure readings for a number of weather stations, fronts and four isobars. The isobars have been drawn to show you how a line of equal pressure can be worked out from the pressure readings given. Make a copy of this map and add the isobars for the following readings: 1016, 1020, 1024, 1028 and 1032.

The map which you have now drawn shows the pressure patterns at ground level. But weather takes place at all levels in the troposphere. The forecaster also wants to know pressure patterns at higher levels. For this he has to use information received from *radio-sondes*. These are miniature weather stations which are sent into the upper air attached to balloons. They send back weather information by radio. Using this information weather men can draw charts which show pressure patterns and other weather features at different levels in the troposphere.

Pressure values reported from land weather stations and ocean-going ships. See the exercise on this page.

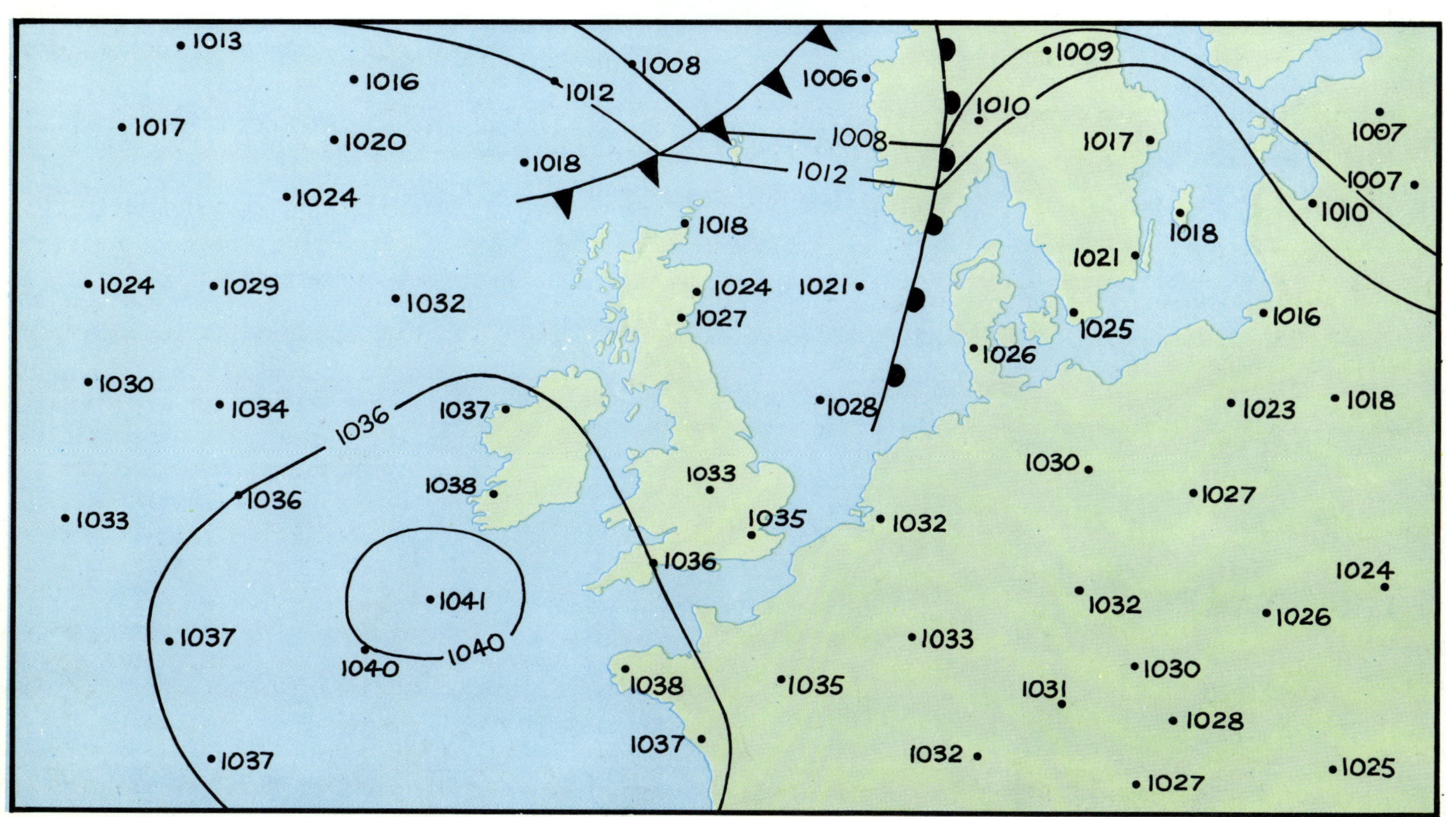

Depressions, or Low Pressure Systems

Polar fronts

Polar fronts occur in two zones that circle the world between latitudes 40° and 65°. They form where tropical air moving towards the poles meets polar air moving towards the equator. The weather systems that develop in these frontal zones usually travel from west to east.

The development of a depression

Frontal zones are not straight lines. They develop waves at ground level. Often they develop a series of waves in quick succession. Each is called a *depression* or *low pressure system*.

On the right there is a series of four weather maps showing how a depression grows. Maritime tropical air meets maritime polar air at a front (1). A small wave forms on this front and the warmer air makes a wedge in the colder air (2). A warm front is seen ahead of a cold front. Near the point where the two fronts meet, an area of lower pressure develops quite rapidly (3). This depression continues to grow. In the northern hemisphere the air moves round it in an anticlockwise direction (4). Remember Buys Ballot's Law (see page 13).

A cross section through a depression

Much of the wet weather that comes with a depression is found along the two fronts. Imagine that map 3 is larger and that a line is drawn across the two fronts. Then a vertical section is drawn through the troposphere. Many depressions would look like the cross section below.

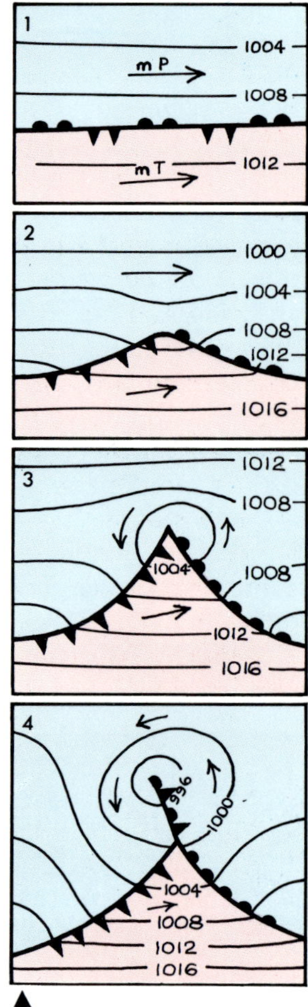

▲ Four stages in the development of a depression

Cirrus clouds may be one of the first signs of an approaching warm front. ▶

In the east, maritime polar air is at ground level. High in the troposphere the maritime tropical air is rising over it. It is cooled and eventually becomes saturated. At the highest levels clouds of ice crystals are formed, mainly cirrus and cirrostratus. These clouds may be seen as much as 1,000 kilometres ahead of the place where the warm front is at ground level.

Where the front is nearer to the ground, the cloud is thicker. Altostratus, stratus and nimbostratus are found. Sometimes there are towering cumulus and even cumulonimbus clouds created by strong up-currents of air. Rain is falling from the altostratus and stratus. The heaviest rain is falling from the nimbostratus.

At the warm front, the direction of the wind changes. It is said to *veer*—that is, to move clockwise. For example, a wind coming from the south will veer to the south-west or west.

Further to the west, the maritime tropical air is found at ground level. This zone is the *warm sector*. Its width varies. Temperatures

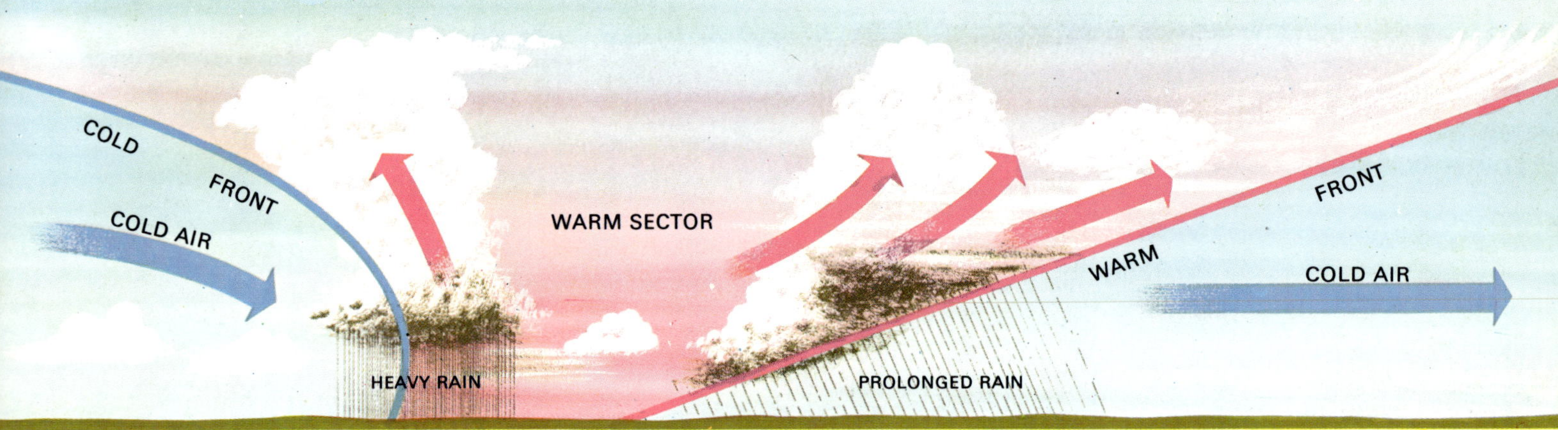

are higher here. The air has a lower relative humidity and there is less cloud. The weather is sunny, but patches of cloud are giving a few showers.

At the cold front, the direction of the wind veers again, this time from south-west or west, to north-west. Eventually, it will veer to north. Here cold air is undercutting warmer air. The warm air is forced above ground level and cools. Another zone of nimbostratus, towering cumulus and cumulonimbus clouds is formed. Rainfall is more heavy than at the warm front, but it does not last so long. Thunderstorms may occur.

To the west behind the cold front, the air is colder than in the warm sector. Cumulus clouds are giving some periods of rain, sometimes called *clearing showers*. Pressure is rising further to the west.

1. Copy the diagram at the bottom of page 22. Look at the pictures of clouds on pages 16 and 17 to help you name all the cloud types present.

The text above describes a depression at one moment in time. As the depression moves from west to east the weather follows the pattern of changes described.

2. On the right are five circles recording the weather at the same weather station at intervals of six hours. Write a brief description of the weather at each time. Then say what weather system you think has passed this station.

Occlusions

Cold fronts generally catch up and overtake warm fronts. When the two fronts join they form an *occlusion* or *occluded front*. Occlusions bring torrential rain over a short period of time.

From the time when a depression first forms until it becomes occluded may be many days. The air ahead of the warm front may not have the same temperature as that behind the cold front. Then the words *cold occlusion* and *warm occlusion* may be used. The word *cold* or *warm* tells you which of the original fronts is still to be found at ground level.

3. Remember that cold air always undercuts warmer air. Copy the three diagrams below. On each one put the following labels where appropriate: warm air, cold air, coldest air, warm front, cold front. Then name each diagram (a) a cold occlusion (b) a warm occlusion (c) a warm and cold front.

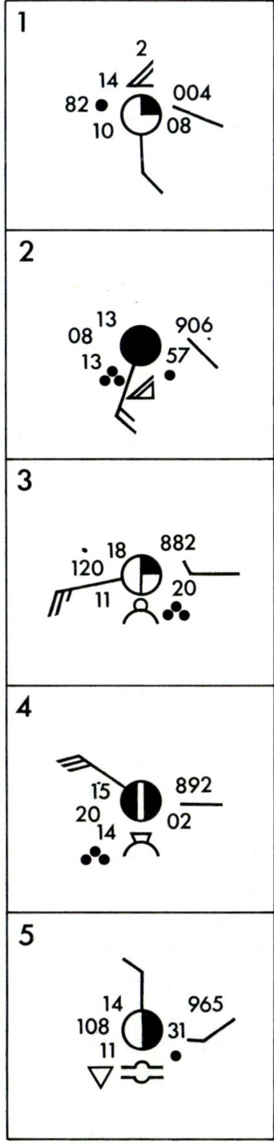

Five station circles showing how the weather changes as a depression passes

Three cross sections to show the main features of different frontal systems

◄ A cross section through a depression

23

Tropical Storms

In tropical areas low pressure systems do not contain fronts between warm and cold air streams. Instead they are made up of giant spirals of rotating air. Some develop into violent storms. One of these spirals is seen on page 30.

1. Using the map on this page, list the countries affected by tropical storms. Write down the name for tropical storms in each country.

Tropical storms form over the sea. Here the intense heat of the tropical sun causes the air to rise. It also causes large quantities of water to evaporate. The water vapour is carried to great heights and creates large areas of thick cloud.

For some reason not fully understood by weather men an area of *intense* low pressure sometimes forms. Around its centre a spiral of very strong winds develops. The clouds in a tropical storm are all of the cumulus type. At the centre of the storm, called the *eye*, there is almost no wind and little cloud.

2. In which direction does the wind in a tropical storm rotate (a) in the southern hemisphere (b) in the northern hemisphere? Look back to page 22.

In tropical storms there are winds which can reach speeds of 200 kilometres an hour. The whole storm may move across the earth's surface at between 30 and 40 kilometres an hour. One storm may be as much as 600 kilometres across.

Because tropical storms are so destructive, man would like to be able to lessen their damaging effects. But no way has yet been found to do so. In some parts of the world a careful watch is kept for them. They are identified early and then tracked closely. In this way their future path can be forecast and some precautions taken.

3. If a tropical storm has been reported in the newspapers recently, collect information about its location, its strength and the damage which it caused. If not, look for descriptions of famous examples in encyclopaedias, old magazines and books.

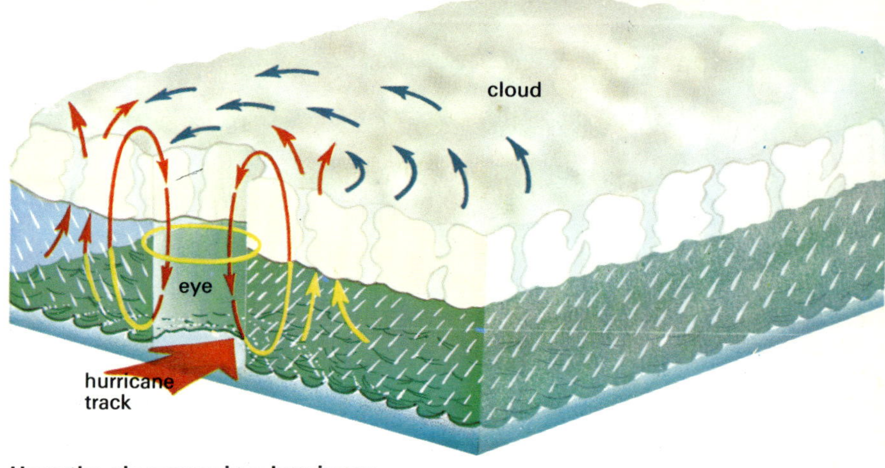

How the air moves in a hurricane

Damage caused by a hurricane in the United States

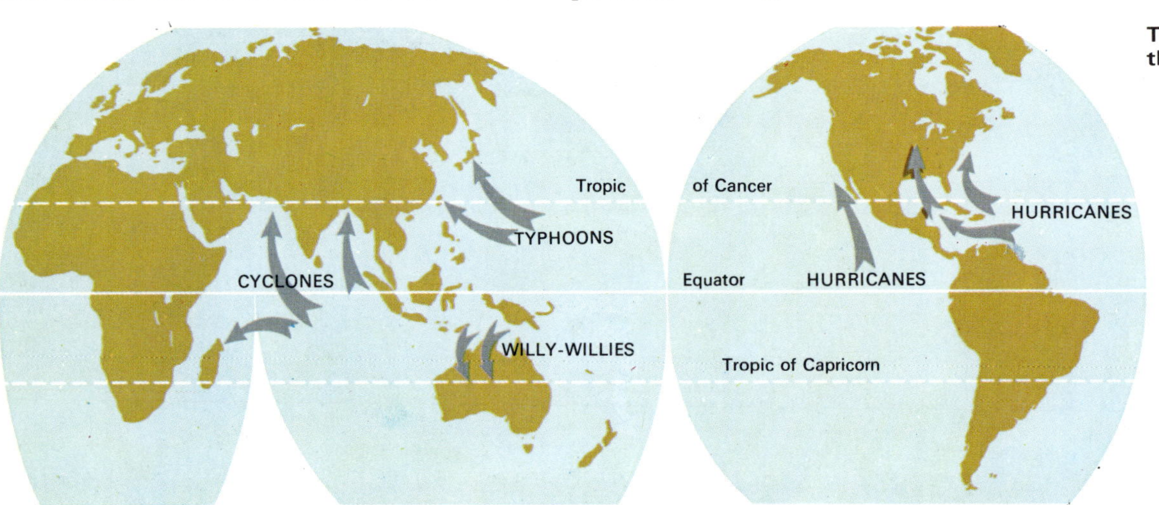

Tropical storms around the world

Anticyclones, or High Pressure Systems

In some places air descends towards the earth's surface (turn to page 5). This creates an area where the pressure is higher than that around it. It is called an *anticyclone*, the opposite of a low pressure area.

Anticyclones generally last longer than depressions. They may last for a week or more, whereas one depression may pass through an area in two or three days. The weather is thus more settled in anticyclones.

Descending air fans out as it moves into the lowest layers of the troposphere. A spiralling motion is set up. In the northern hemisphere, at the earth's surface, the winds move in a clockwise direction around the centre of an anticyclone.

The isobars in an anticyclone are usually far more widely spaced than they are in a depression. As explained on page 13, the winds are therefore far lighter. In some

▲
Radiation fog at an airport. In such conditions airlines may be brought to a halt.

A January day on the Spanish Meseta. Still, anticyclonic conditions bring temperatures of −5°C and low relative humidity. ▶

A weather map showing an anticyclone over the British Isles
▼

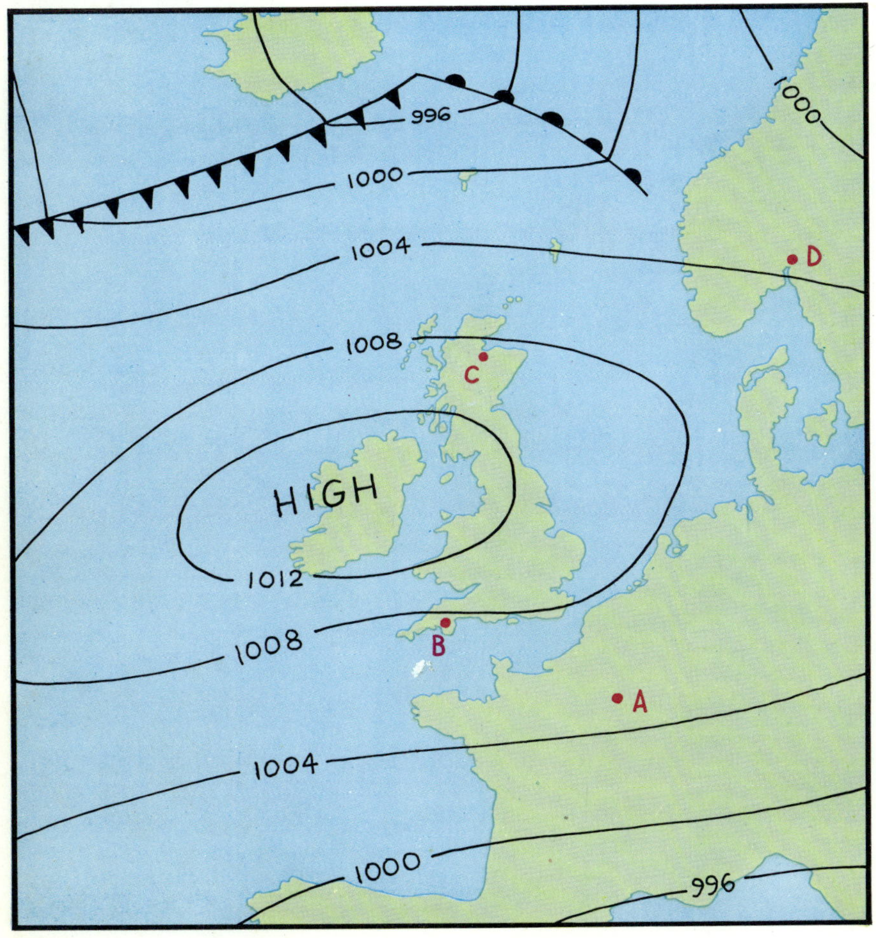

places an anticyclone brings calm conditions with no wind at all.

Descending air warms up. In an anticyclone there is therefore little chance of cloud developing. This results in very different weather conditions in summer and winter. With clear skies the sun should shine by day, but heat will be quickly lost at night. In summer, daytime temperature may be high, but at night they may drop quite sharply. In winter, when the sun is low in the sky, daytime temperatures will not rise much. At night, they will fall rapidly and often create ideal conditions for the formation of frost and radiation fog.

1. On the left is an outline weather map with a large anticyclone off western Europe. Look back to pages 18 and 19. Then make two copies of this map, one for December, the other for June. Draw four station circles to show what wind speed, wind direction, cloud cover, air temperature and weather the four marked weather stations may each be reporting (A—Paris; B—Plymouth; C—Inverness; D—Oslo). Then write brief descriptions of the two weather charts.

2. One photograph on this page shows how fog can be a hindrance to man. Make a list of other ways in which fog affects him.

Thunderstorms

Thunderstorms sometimes occur during an anticyclone. On a hot summer's day air at ground level in some places may be heated rapidly. It will then rise and may take with it large amounts of water vapour. If the air rises to a great height the clouds which form range through cumulus and towering cumulus to cumulonimbus.

Such clouds are very turbulent and have strong up and down-currents within them. In these currents ice crystals and rain drops are often broken up. At the same time, electrical charges are separated out. At the top of the cloud a *positive* charge is created. At the base of the cloud a *negative* charge is created.

When the build-up of charges is large enough, the electricity is discharged as·a flash of lightning. The flash jumps between the top and bottom of one thunder cloud,

A tropical storm over Darwin, Australia

between one thunder cloud and another, or between a thunder cloud and the earth beneath it. On the ground, high trees and buildings are most likely to be struck.

When the lightning passes through open spaces it looks like a crooked, branched line. This is called *forked lightning*. When the discharge is inside a cloud or is masked by a cloud it appears as a broad flash of light. This is called *sheet lightning*.

1. What damage can be done where lightning strikes? What is the purpose of a lightning conductor?

2. If you are caught in a thunderstorm in the countryside where should you not shelter? Why?

As lightning passes through the air intense heat is created in a small area. The air expands and contracts very quickly. This produces a sharp sound, called *thunder*. It rumbles because the original sound echoes inside the clouds and from the earth's surface.

3. Light travels at a speed of 300,000 kilometres a second, and sound at only 350 metres a second. For this reason, thunder always follows a flash of lightning. They occur at the same time only when the thunderstorm is directly overhead. Work out a formula from which you can tell the distance that you are away from a thunderstorm.

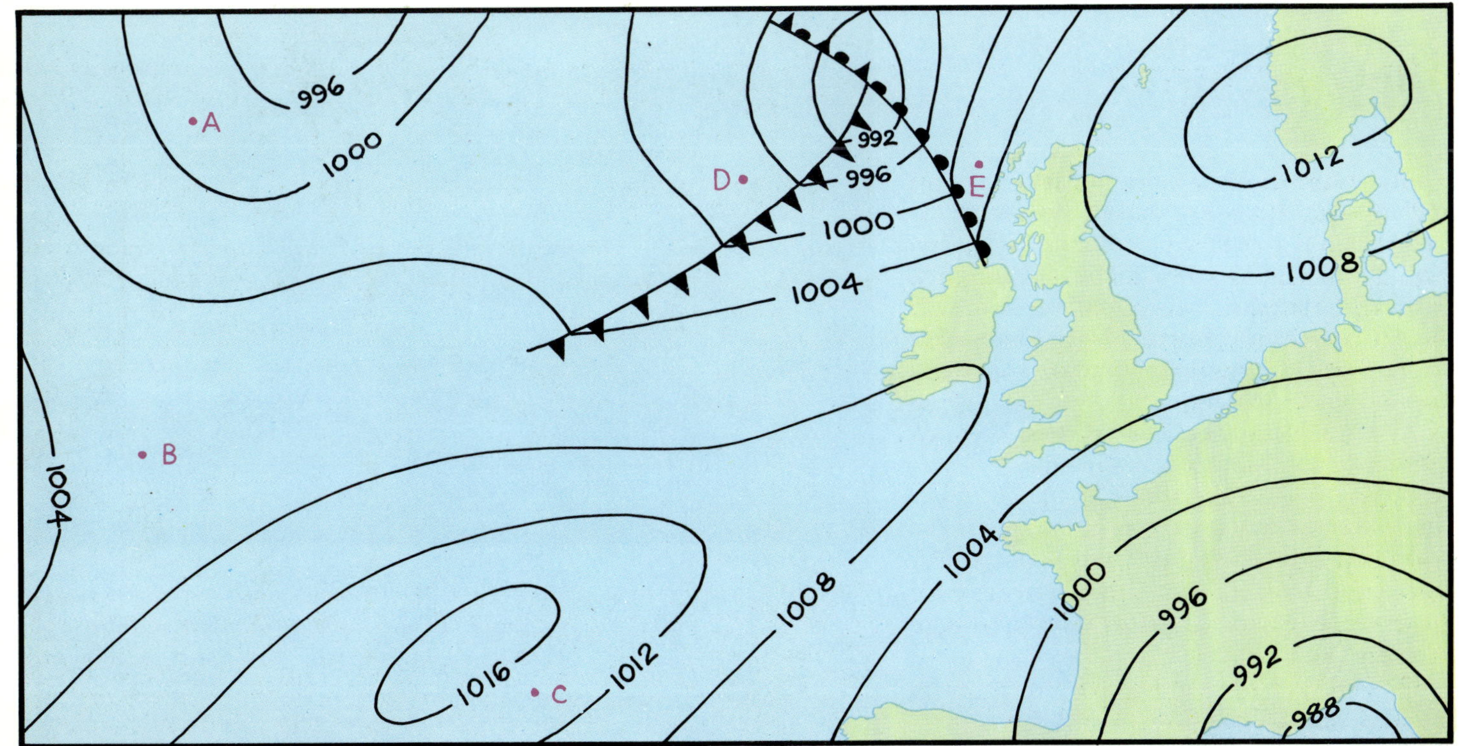

Troughs, Ridges and Cols

Above: a complex weather situation over the Atlantic Ocean and north-west Europe

A trough of low pressure

At the centre of a depression the isobars completely surround the area of lowest pressure (see pages 22 and 23). There is another low pressure system in which the isobars are arranged in an open V shape. The pressure is lowest in the middle of the V. This pattern is called a *trough* of low pressure. On a weather map it may be shown by dashes drawn across the isobars.

Air moving into a trough comes together from different directions. Some air is forced to rise. This air then cools and a belt of cloud is formed. The weather that results is a period of heavy rain.

Troughs of low pressure often develop in groups. A forecaster in western Europe may then talk of 'a series of troughs approaching from the Atlantic bringing belts of rain with clearer periods in between'.

A ridge of high pressure

At the centre of an anticyclone the isobars completely surround the area of highest pressure (see page 25). Often an anticyclone develops an extension in one direction. This has a V-shape, with the highest pressure in the middle. This is called a *ridge* of high pressure.

Air moves away from the centre of a ridge, and descends. The weather associated with it is similar to that of a complete anticyclone. In summer it is usually fine and sunny. Clear weather occurs in the winter, perhaps with fog or frost.

A ridge of high pressure often forms between two depressions. It brings a short, settled spell between two periods of rainy weather.

A col

Some synoptic charts show an area which has neither high nor low pressure. The isobars show no set pattern and are usually wide apart. There are areas of high and low pressure, sometimes two of each, and their edges meet in this one area. Winds here are described as *light and variable*. The weather is unpredictable. This pressure pattern is called a *col*.

1. Above is a weather map showing pressure patterns and some fronts. Copy this map and write the following words in the correct places: low (twice), high (twice), ridge, trough, col (twice), warm front, cold front, occlusion.

2. On your map fill in the wind speed and directions which would be reported by ships in the Atlantic Ocean at the places marked A, B, C, D and E.

Synoptic Charts and Weather Forecasts

Collecting weather information

When all the elements that make up the weather have been measured at a weather station the details are exchanged at national level. Weather measurements are exchanged four times every day at 00.00 (midnight), 06.00, 12.00 and 18.00 hours Greenwich Mean Time.

At these times records of temperature, pressure, winds, clouds and so on are made at hundreds of land stations, special weather ships and other ocean-going vessels. The information is then quickly transmitted by radio and teleprinter. It would take a lot of time to exchange such information in words. Therefore a series of special codes is used. Here is an example: 772 61812 80011 27904 00901 51722. The first part of this means: 722—London (Heathrow) reporting; 6—6 oktas of cloud; 18—the wind is blowing from 180°, or due south; 12—the wind speed is 12 knots.

The other groups of five figures report the visibility, present weather and past weather; pressure and dry bulb temperature; types and heights of clouds; dew point temperature and the way in which pressure has been changing in the last three hours.

Each weather station sends its records to a regional centre. In turn, this passes the records to the national centre, such as the British meteorological office at Bracknell

A weather man plotting a synoptic chart

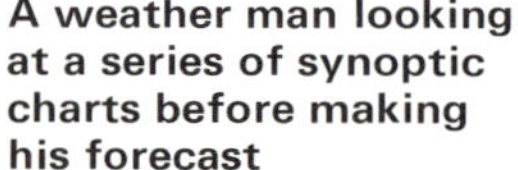

A weather man looking at a series of synoptic charts before making his forecast

in Berkshire. Finally, each national centre exchanges its country's weather record with that of neighbouring countries. In this way a complete record of the weather over a large area is available very soon after all the station reports have been made.

Drawing a synoptic chart

The photograph above shows a weather man drawing a synoptic chart from a collection of coded records. The weather man first draws circles like those on page 6. Next, he identifies and plots the positions of fronts. Finally, he fills in the pressure patterns. Until recently this was the standard way of drawing a synoptic chart. Today, it is becoming more usual to have the chart printed out by a computer at each national centre.

Making a forecast

A synoptic chart provides a record of the weather at ground level at a stated time. When a series of charts is drawn at intervals of 6 hours it is possible to see how the weather is changing.

Other charts are drawn to show how weather elements are changing at various levels of the upper air.

By using all these charts the weather man can make a detailed forecast of the weather in the next 24 or 48 hours. He can also make a more general forecast of the weather during the following days. The photograph on the left shows a senior forecaster carefully studying a series of synoptic charts. He is considering the ways in which the weather systems are moving and the weather elements changing. It is now possible for a computer to print out a forecast chart.

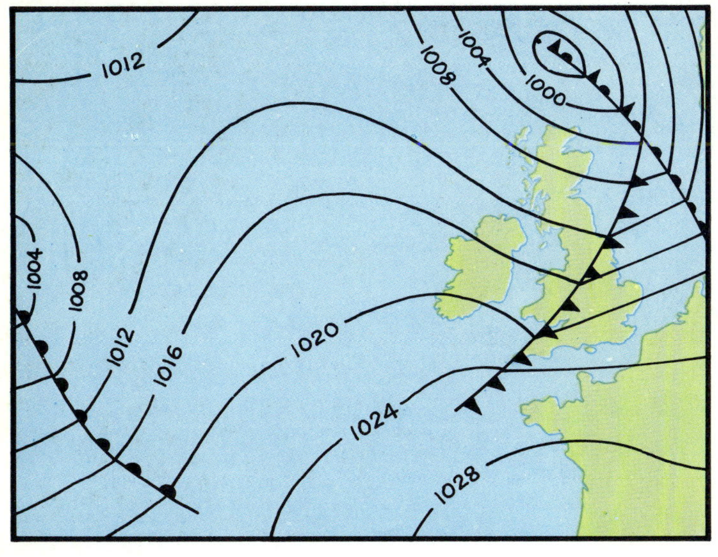

1. Synoptic chart for midnight

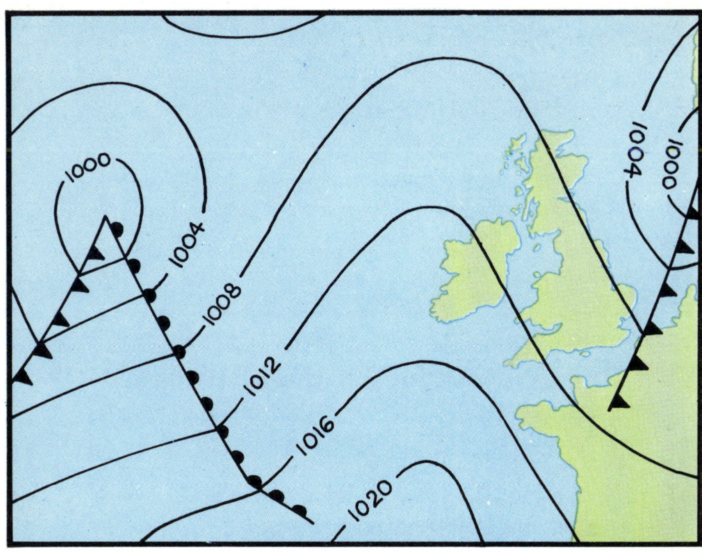

2. Synoptic chart for 06.00 hours

Suppose that you are a forecaster studying the series of maps on this page. The maps are drawn at intervals of six hours.
1. What are the main air streams in the area?
2. What fronts are present? How quickly are they moving?
3. What are the main pressure patterns found on the maps? Are any anticyclones or ridges stationary or moving?

The four maps on this page are synoptic charts recording weather readings at different times on a day in March.

4. If the weather systems continue to move at the same rate during the next 24 hours, where will they be at the end of that period? Make a sketch map to show the possible pattern of isobars at that time.
5. Using your answers to the questions above, forecast what sort of weather may be expected over the area on the map during the next 24 hours and the 24 hours after that.

3. Synoptic chart for 12.00 hours

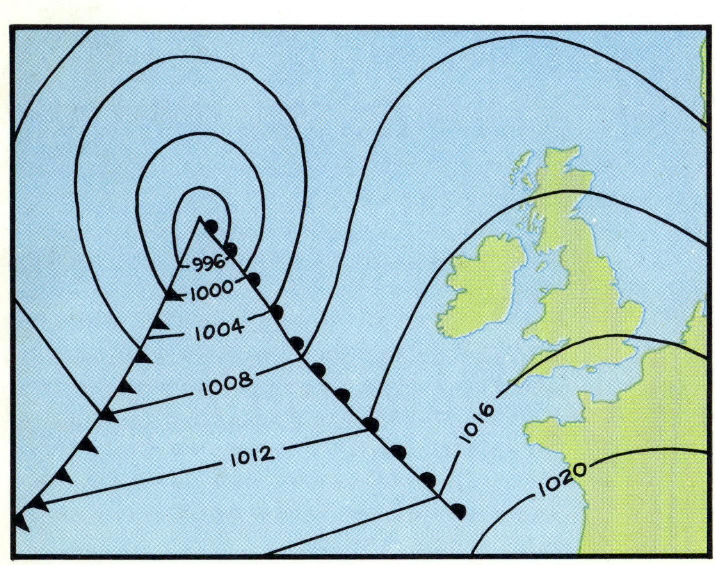

4. Synoptic chart for 18.00 hours

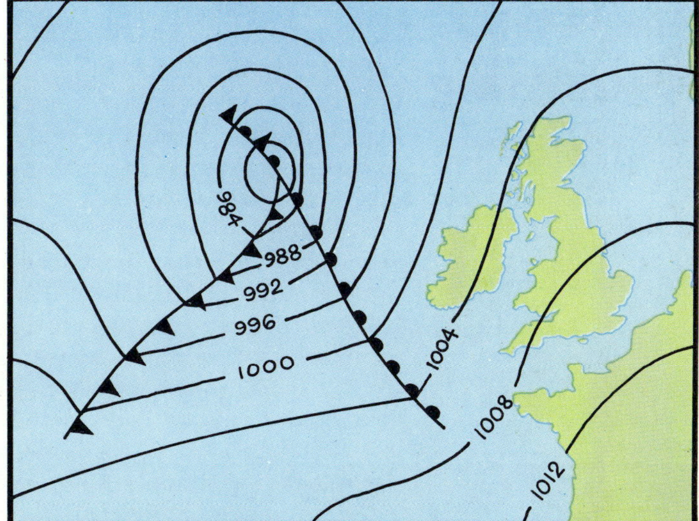

Advances in Forecasting Techniques

In recent years a number of technical advances have helped weather men to gather and interpret weather information. Some enable them to handle a lot of information more quickly. Others provide information which was previously not available. Here are a few of these new advances.

Satellites

Weather satellites orbit the earth at a height of about 700 kilometres. Their great service to weather men is to provide photographs of the clouds and weather systems in the earth's troposphere. These photographs are radioed back to the earth. They provide information even for areas where there are no ground stations.

Photographs similar to those taken from weather satellites are often taken by astronauts in manned spacecraft. Above is a photograph of hurricane 'Gladys' taken in 1968 over the Gulf of Mexico. It clearly shows the spiral pattern of the clouds typical of hurricanes (see page 24). Satellite photographs such as this are a great help in understanding these important features of the weather.

1. Make a sketch weather map to show the positions of the main pressure systems and weather features seen in the photograph.

2. Look in the newspaper to find out when a weather satellite can be seen at night. This information is often printed beside the weather map and forecast. Then watch the satellite cross the sky.

Radar

Radar is a device which shows on a screen the direction, size and distance of an object. It does this by bouncing short radio waves off it. In weather forecasting, radar is used to find the positions of showers, rain belts, thunderstorms and so on.

Some radar screens look like television sets. The centre of the screen is the position of the observer. Distances and directions are measured from that spot. On other screens the position of the observer is in one corner of the screen. Distances and heights of objects are measured from that spot.

3. The two pictures on the right show a belt of rain and clouds as they appear on radar screens. The picture above shows a plan view, the other shows a vertical section. In what direction is the belt of rain from the observer? What is the thickness of the cloud? How far away is the rain falling?

Computers

Computers handle millions of weather observations each day. They can also draw synoptic charts and forecast charts for short periods. In addition they can store all previous weather records. The latest pattern of weather can therefore be compared with patterns in the past. When the closest comparison is found long-range forecasts can be made.

4. Study a number of daily weather forecasts and any long-range forecast. Compare the detail which they give.

5. Study the latest long-range forecast. See how accurate it is during the period to which it applies. Do this exercise several times over a longer period of time.

Automatic weather stations

In many parts of the world it is not difficult to run manned surface weather stations. In other parts few people live, and manned weather stations are very costly and difficult to run. Today more and more automatic stations are being set up. These provide information in such places as polar regions and deserts. Seas through which few ships pass have anchored buoys to which automatic stations are attached. The radio-sonde is another example of an automatic weather station (see page 21).

6. The radio-sonde balloon bursts in the upper troposphere. The instruments fall back to the earth. Find out how the national meteorological office gets them back.

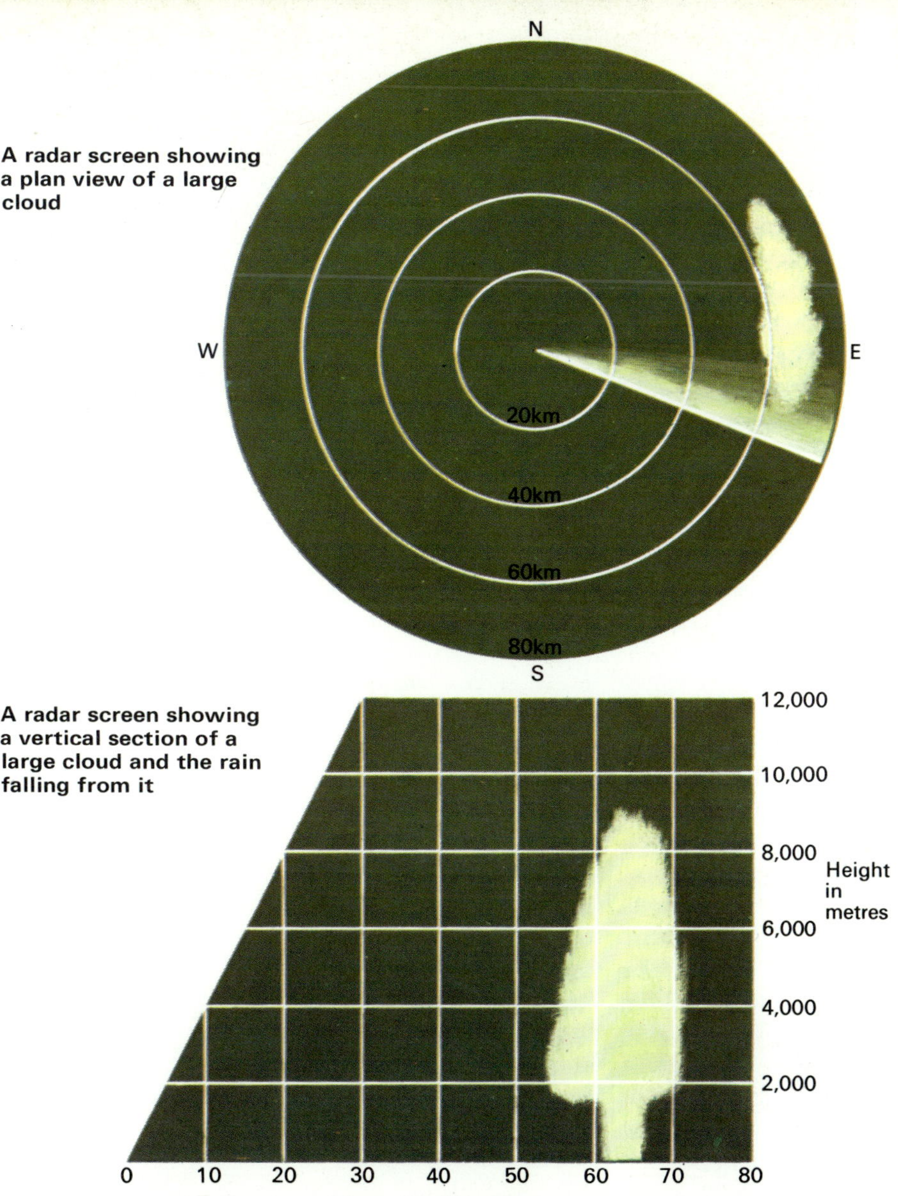

A radar screen showing a plan view of a large cloud

A radar screen showing a vertical section of a large cloud and the rain falling from it

A weather chart which has been printed by a computer being taken out ▼

Summary Questions

1. The photograph on the right shows three different types of cloud. Draw simple pictures to illustrate each, giving them names. Write a short sentence to describe the height and characteristics of each cloud type.

2. The maximum and minimum thermometers on the right show a setting which you might find at 09.00 hours one morning (a) Look at the minimum thermometer. What is the temperature as you look at the thermometer? What was the lowest temperature during the previous night? (b) Look at the maximum thermometer. What was the highest temperature during the previous day?

3. Make a table which matches the following types of tropical storms with the countries in which they may occur: hurricane, cyclone, typhoon, Willy-Willies. Australia, India, U.S.A., Eastern China, Bangladesh, Malagasy, New Guinea, British Honduras, Japan.

4. You see a flash of lightning and then hear the accompanying clap of thunder. How many kilometres are you from the thunderstorm if the time difference is (a) 10 seconds (b) 3 seconds (c) 27 seconds?

Clouds over a lake in Maine, U.S.A., on an August evening

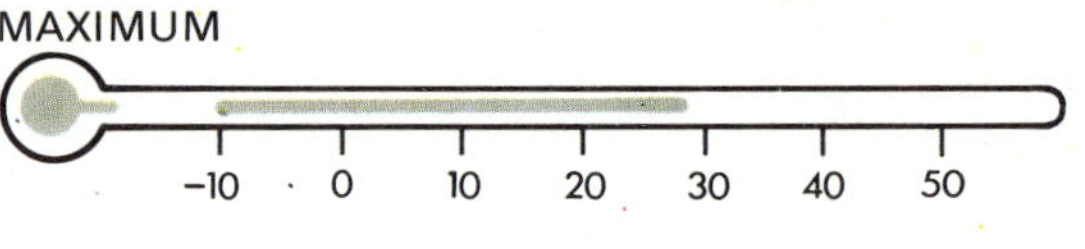

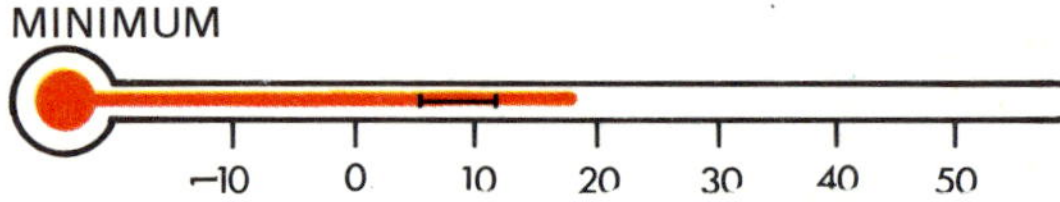

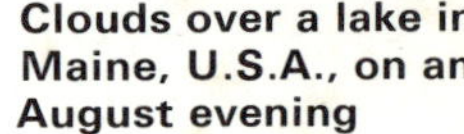

5. What does the station circle on the left tell you about the following weather features: wind speed, wind direction, the amount of cloud, the dry bulb temperature, the wet bulb temperature, the pressure tendency, and the types of cloud?

6. What do you think the following weather sayings mean?

(a) 'Rainbow to windward, foul falls the day. Rainbow to leeward, damp runs away.'

(b) 'The north wind doth blow and we shall have snow.'

7. The graph on the left shows the rainfall recorded each day in April, 1972 in Glasgow. When was the wettest seven-day period? How much rain fell then? How many days had no rain? How long was the longest dry spell? Which was the wettest day?

8. Describe the pressure changes recorded on the barogram printed below. Refer to the readings in millibars.

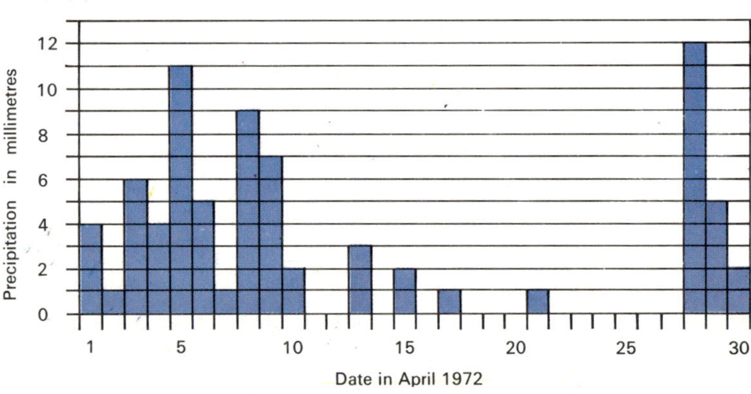

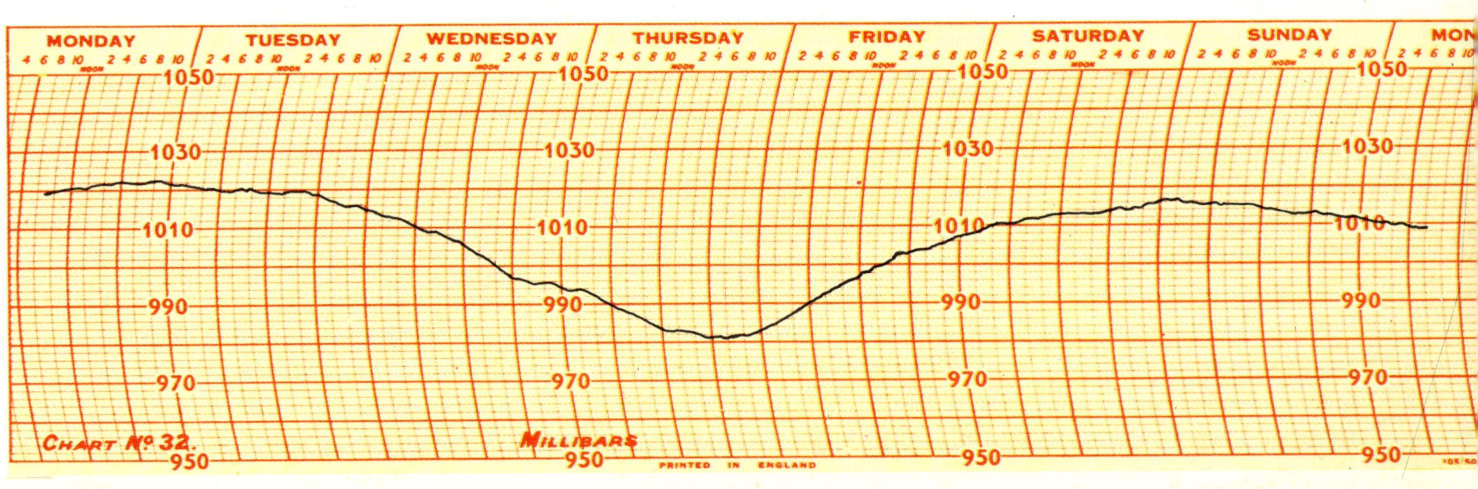